Getting Started
with the
TI-89 Graphing Calculator

Carl Swenson

Seattle University

WILEY
JOHN WILEY & SONS, INC.

To order books or for customer service call 1-800-CALL-WILEY (225-5945).

ISBN 0-471-74206-6

10 9 8 7 6 5 4 3 2 1

PREFACE

The purpose of this book is to show how to apply the features of the TI-89 and Voyage 200 graphing calculators to understand calculus. The book is divided into five parts, corresponding to common areas of focus in a calculus course. The chapters provide a more specific description of each calculus topic. In general, if you are looking for help on a calculus topic, then use the Table of Contents to find the topic, but if you are looking for help on a calculator command, start by looking in the Index. Each calculus chapter is intended to be stand-alone but they all require an understanding of the basics from Part I Precalculus. Part I is intended as a review; it can be skimmed by experienced users or used as a primer by new users of this calculator.

I would like to acknowledge and thank Deborah Hughes-Hallett and the Calculus Consortium for Higher Education (CCHE) for permission to use examples from their work.

To the student

Using a graphing calculator can be both frustrating and fun. A healthy approach when you get frustrated is to step back and say, "Isn't that interesting that it doesn't work." Figuring out how things work can be fun. If you get too frustrated, then it is time to ask a friend or the instructor for help. Make sure you have a phone list of friends with the same calculator.

Part I gives you clear sets of key sequences so you become comfortable with how your calculator works. The remaining parts shift into a higher gear and only show you calculator screens as guides for the keystrokes. Your *TI Guidebook* provides a resource if you get stuck; it explains each feature briefly, usually with a key sequence example.

Remember that the *Guidebook* is like a dictionary: there is no story line or context. In this book, the features that you need for calculus are explained in the context of calculus examples. Other calculator features that are less important to calculus may not be mentioned at all. The mathematical content drives this presentation, not the calculator features.

I have included tips about such things as short-cuts, warnings, and related ideas. I hope you will find them useful.

Tip: Don't use technology in place of thinking.

To the instructor

These materials are designed to allow you to focus on the calculus, not the calculator. By having the students use a single calculator specific book, you should be able to greatly reduce the problems caused by using multiple calculator materials.

Will these materials take care of all your students? Of course not. There will still be the zealous ones who want the programs in assembly language and the anxious ones who want the buttons pressed for them. These materials are aimed at the middle, giving enough guidance so that most students are able to work through an example without assistance, but not so specific as to be considered a mindless exercise in pressing keys in the right order.

Programming is not an emphasis of this book. I have included five programs which I feel enhance the calculus learning. Find a techno-hungry student to enter them and insure that they are running properly. Then distribute them to your class using LINK.

Tip: The TI Volume Purchase Plan provides you with a classroom calculator and/or an overhead model for classroom use.

Dedication

This book is dedicated to all golden retrievers. They know the calculus of minimizing the distance to a frisbee and maximizing the fun.

Carl Swenson
swenson@seattleu.edu

TABLE OF CONTENTS

PART I PRECALCULUS

PART I PRECALCULUS (continued)

PART II DIFFERENTIAL CALCULUS

PART IV SERIES

PART V DIFFERENTIAL EQUATIONS

Notes:

CHAPTER ONE

GETTING STARTED

The tools used to make numeric calculations have developed from the fingers, to an abacus, to a slide rule, to a scientific calculator, and now to a calculator capable of symbolic manipulation. In this chapter we see how to use the TI-89 — a current tool of calculation. If you have used a graphing calculator before, you may only need to skim this chapter to understand the differences between your old and new calculator. This chapter is a brief summary of that material. In this book the references to TI calculator keys and menu choices are written in the TI font. The TI font looks like this.

A note about the TI-89, TI-92 and Voyage 200

The screens shown in this book are from the TI-89, however the basic computations and commands in this book are the same as those on the TI-92, TI-92 Plus and Voyage 200. The older TI-92 (non-Plus) does not support features for differential equations. The screens of the TI-92 and Voyager are similar but larger than those shown from the TI-89. The TI-89 has downloadable updates for its operating system; version 3.0 was used to produce this book.

Getting started with basic keys

The ON key
Study the keyboard and press the ON key in the lower left-hand corner. You should see a blinking rectangular cursor. If not, then you may need to set the screen contrast. Even if the cursor is showing, it is a good idea to know how to adjust the screen contrast.

Using the ♦ key to adjust the screen contrast
As you use the calculator, the battery wears down and it becomes necessary to adjust the screen. Also, you may need to adjust the screen contrast for different lighting environments. Press and release the ♦ key and then the + (plus) key in the upper right of the keypad. By repeating this sequence, the screen darkens. The screen lightens by repeating this sequence but by using the − (minus) instead of the plus key. If the setting is too low, the cursor does not show; and if it is too high, the screen is dark as night.

If you take a break and come back later, the cursor disappears for a different reason. The calculator goes to sleep; it turns itself off after a few minutes of no activity. Just press the ON key and it wakes up at the same place it turned off: no memory loss.

Tip: Sometimes "broken" calculators are fixed by reinserting the batteries correctly.

The HOME screen

The default screen as you turn on your calculator for the first time shows the APPS menu. Select Home, it is on the home screen that we work. On most graphing calculators, results flow down the screen as you work. These models of the TI work differently. Commands are entered from a lower entry line and answers scroll up the screen into the history area. Look for the highlighted item or the blinking cursor to know where you are. The four areas on the HOME screen from top to bottom are:

- *Toolbar* — Top row: pull down menus activated by the corresponding function keys, F1 to F8, located just below the screen.
- *History* — Main screen: displays previous instructions and answers. Use up-arrow to scroll and highlight instructions or answers. They can be copied to the entry line.
- *Entry line* — Action line: input and edit expressions or instructions.
- *Status line* — Bottom row: current settings and messages in fine print.

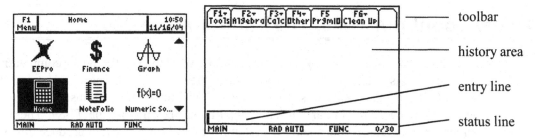

Figure 1.1 The Apps Desktop, when active. The HOME screen area is always available by pressing the HOME key.

Tip: The HOME key is an often used key. When you are lost go HOME.

Entering expressions and instructions

On the right side of the calculator are the arithmetic keys used on a scientific calculator. Master these first. The numeric keys are used for simple arithmetic calculations. In this chapter, the keystrokes are given for the results shown on the screens. In later chapters, the screen speaks for itself.

Tip: Test technology with known results before trying complex examples.

We start by typing

$$3 \div 2 \text{ ENTER}$$

Be aware that the symbol on the divide key, ÷, is different from the divide symbol (/) on the screen. The screen in Figure 1.2 shows the input on the left and output on the right. Notice that the integer division was not done. Unlike most calculators, the TI responds with the exact form instead of giving a decimal answer. (We shortly learn how to put answers in decimal format.)

> *Tip:* Some keypad symbols, like +, are displayed on the screen with a different symbol from that on the key.

Make special note that the (-) key is for negation and must be distinguished from the – subtraction key. One of the most common errors is interchanging the use of the subtraction key and the negation key. Give it a moment's thought: subtraction requires two numbers, while negation works on a single number. Try pressing the following sequences of four keys.

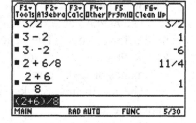

$$3 - 2 \text{ ENTER}$$

and then

$$3 \ (-) \ 2 \text{ ENTER}$$

Figure 1.2 Division symbol, and subtraction vs. negation key distinction.

Look carefully at Figure 1.2 which shows the subtraction symbol is longer and centered, while negation is shorter and raised. The first expression gave 1 and the second entry gave -6 as an answer. By looking at the screen you see a multiplication dot between the two numbers to show you that it multiplied 3 times -2 when you entered 3 (-) 2. An important feature of this calculator is that it formats both input and answers, a feature called *pretty print*.

> *Tip:* Before using an answer, check the input format in the history area to see that you entered the expression correctly. The expression is still highlighted on the entry line and can be easily edited and recalculated if necessary.

The single most popular error (can errors be popular?) among new users is to not use parentheses when needed. This is a serious error because the calculator does not stop and alert you with an error screen; instead, it gives you the correct answer to a question you are not asking. Suppose you want to add 2 and 6 and then divide by 8. You don't need a calculator to tell us the expression has value 1. But if you enter 2 + 6 / 8, the given answer is 11/4. The calculator divided 6 by 8 first and then added that to 2. Use parentheses to insure that you are evaluating the correct expression. Now try (2 + 6) / 8 and get an answer of 1 as expected. Note how the expression is shown in the history area as a fraction. There is a prescribed order of operations on your calculator and the pretty print feature shows an unambiguous expression and its answer.

> *Tip:* When you get an unexpected result, go back and check parentheses. Be generous; adding extra parentheses doesn't hurt.

Magic tricks to change the keyboard: 2nd, ♦, ALPHA, ↑

To do more with limited keys use the modifier keys, they are like Shift and Ctrl on a computer keyboard. A difference from a computer keyboard is that the modifier keys on the calculator are pressed once and not held down. The indicator of a modified condition is on the status line just to the right of the word MAIN. The first modifier key is the 2nd key. After pressing it once – presto! – all the keys now have a new meaning. These meanings are indicated just above each key on the left. The notation 2nd_SIN is used in this book to denote that SIN is an entry that needs to be first modified by the 2nd key.

We have already used the ♦ key to adjust the screen contrast. It also provides easy access to graphs and tables when used with the top row of function keys. Its entry action is shown on the right above the top of the key for the upper four rows. Pressing ♦ and then the up-arrow moves the cursor to the beginning of the entry line; using the down-arrow instead moves the cursor to the end of the line.

The ALPHA key provides the alphabetic letters (a-z) as shown on the right above the keys in the lower rows. The space key has a symbol ⌴ and is above the negation key. Press 2nd_ALPHA and the cursor remains in alpha mode. This is handy for entering long alphabetic commands or text. To release the alpha lock, press ALPHA once.

The fourth modifier key is the ↑ key. It is used like a shift key on a computer to enter uppercase letters.

> *Tip:* The color of the 2nd, ♦, ALPHA and ↑ keys helps distinguish the modifier key action. The colors used on different models of calculator vary.

> *Tip:* The 2nd, ♦, ALPHA and ↑ keys work as toggles: if you press one by mistake and turn on a modifier, just press the key again to turn it off.

The greatest equation ever written

There are five symbols, 0, 1, e, π, i, that are frequently used in mathematics. Incredible as it might seem, they can be related by a single equation,

$$e^{i\pi} + 1 = 0.$$

Practice using the modifier keys for e, π and i by entering

♦_e^x 2nd_ i 2nd_ π) + 1 ENTER

Notice that the symbol e^x on the key displays as e^(on the input screen and then as plain e in the pretty print history area. Exponentiation is written in mathematics as a superscript; this is easy by hand, but not directly possible on a calculator. The caret ^ is used to signify exponents. The complex i is not the lower case i. The complex i is above the CATALOG key.

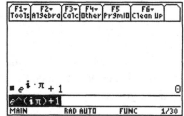

Figure 1.3 Equation e^(iπ)+1=0.

Getting around: navigation and editing keys

We all make mistakes; correcting them on a graphing calculator is relatively easy. The basic navigation device is the set of four arrow keys.

Correction keys: ←, 2nd_INS, ♦_DEL and CLEAR

When you entered an expression and pressed ENTER, the input remained highlighted on the entry line. Pressing any key other than an arrow deletes a highlighted expression. But pressing the left or right arrow unhighlights the entry line and allows you to edit the expression from the beginning or end, respectively. Insert is the default mode for corrections and the cursor shows as a blinking vertical line. This means that any key you press inserts the corresponding symbol at that point in the expression. Press the backspace key (←) to delete a character to the left of the cursor. Press 2nd_INS and the cursor changes to a blinking square indicating that the correction mode is now overwrite. Press ♦_DEL to delete one character to the right of the cursor. Use CLEAR to delete all the characters to the right of the cursor. Pressing CLEAR twice clears the whole line.

> *Tip:* To clear the entire history area, press the F1 key and then 8:Clear Home. (Pull-down menus such as F1 are explained ahead.)

Deep recall: 2nd_ENTRY

Often you see errors to correct after entering a line or you would like to incorporate previous results into a new expression. You can access not only the previous entry line, but anything in the history area, even entries that have scrolled off the screen.

For example, the pitch of a musical note is determined by the frequency of its vibration (measured in hertz). Middle C vibrates at 263 hertz. The frequency of a note n octaves above middle C (use negative numbers for octaves below) is given by $V = 263{\cdot}2^n$. To find the frequency two octaves higher, enter 263*2^2, see Figure 1.4. To find the frequency two octaves below, edit the first entry and insert a negative sign before the 2 in the following manner. With the previous entry highlighted on the input line, press right arrow to unhighlight it, left arrow to move before the 2 and press (-) to insert a negative sign.

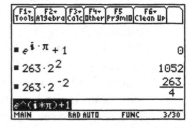

Figure 1.4 Using the recall entry and the insert mode to find two similar calculations.

In addition to having the last line available, by repeatedly pressing 2nd_ENTRY key, all the previous entry lines in the history area are accessible. This is called deep recall. In Figure 1.4 after the octave calculation the greatest equation has been recalled to the entry line. This method erases anything previously on the entry line and cannot be used for inserting — it is all or nothing. However, to insert a result (or a combination of results) from the history area, use the up arrow which takes you to the history area and press ENTER to insert the highlighted expression into the entry line. And finally (as if we need yet another way), this calculator supports the common word processing commands for cutting (♦_X), copying (♦_C) and pasting (♦_V) a highlighted expression. To highlight any set of characters, hold down the shift key (↑) and arrow over the desired characters. This is the rare exception to the rule that modifier keys are pressed and released.

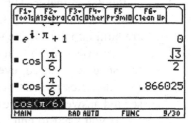

Figure 1.5 Using ♦_≈ to convert exact answers to approximations.

The format of numbers

Most calculators show decimal results, but the TI-89 gives exact answers whenever possible. This is sometimes inconvenient and you may want to change the format of the answer.

Fractions and decimals: ♦_≈

You may have been surprised by the previous answer of 263/4 being shown as a fraction. In Figure 1.5 we evaluate cos(π/6) and get an exact form answer, next we press ♦_≈, (look above ENTER,) to translate the answer to a decimal approximation.

Scientific notation: Folding a paper to reach the moon

If you use really big numbers, they display in full accuracy whenever possible. Let's try an unrealistic but surprising situation that uses a big number.

If you fold a piece of paper, it doubles in thickness. You can measure the thickness in sheets: one fold has a thickness of 2 sheets, two folds 4 sheets, three folds 8 sheets, etc. The formula for doubling is $S = 2^n$, where n is the number of folds and S is the number of sheets.

Verify the astonishing fact that it takes only 42 folds to reach the moon from earth. (We should mention, though, that it is physically impossible to fold a single sheet more than about seven or eight times.)

Enter 2^42 and find the answer in number of sheets in an exact integer form. See Figure 1.6. But long strings of digits are often hard to immediately see as billions or trillions. Press ♦_≈ and the answer is shown in scientific form. The exponent 12 gives us the magnitude. We have an answer in trillions (4.39805E12 means 4.39805×10^{12}). Notice that the exact answer appears to be rounded to six significant digits but, in fact, all the decimal accuracy is retained, as we see it used in the next step.

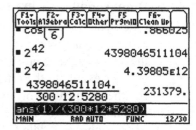

Figure 1.6 Folds to the moon. Large numbers can be forced into scientific notation.

Now to verify that this stack reaches the moon, we convert this number of sheets to miles. On the entry line of Figure 1.6, you see the expression ans(1): this was not typed, but was automatically pasted there when we pressed the divide key at the start of the calculation. If you start an expression by pressing an operation key (+, -, *, /, ^), the calculator assumes the first number in the calculation is the previous answer. We divide our number of sheets by 300 (an approximate number of sheets per inch) and by 12 (inches per foot) and by 5280 (feet per mile). Be certain to put this product in parentheses so the calculator knows what you intend as the denominator. This conversion to miles shows that the thickness of the folded paper is more than the distance from the earth to the moon, about 230,000 miles. (Unit conversions are built into the TI; see your *Guidebook*.)

To insert ans(1) anywhere in an expression, press 2nd_ANS. Using ans(1) insures that full accuracy is used, even if the previous answer was shown in an approximate format. The (1) identifies the first answer above the entry line in the history area. Thus, ans(2) is the second answer above the entry line, etc.

If you want to enter five billion without entering all those zeros, you can use the EE key to insert the E symbol. Press

5 EE 9 ENTER

The screen result is not shown here, but by pressing EE a single E is shown on the screen. Beware that using the capital-E does not work.

Tip: In the history area of the screen, a number whose absolute value is less than 0.001 is displayed in scientific notation unless the default settings are changed.

How to put yourself in a good MODE

The MODE setting choices are described in the TI *Guidebook*. We mention settings as we need them, but unless noted otherwise, all our examples assume that the default settings are in effect.

Numerical format

You can control the output format of numeric calculations so that they are all shown in scientific notation. Or, if you are doing a business application, you might want money answers to come out rounded to two decimal places for the dollar and cents format. Pressing the MODE key allows you to check and change formats. See Figure 1.7.

To change a setting, use the down arrow to reach the desired line, then use the right arrow to display the submenu of settings. Arrow to the one you want and press ENTER to make the selection. Pressing ENTER again returns you to the HOME screen and saves your changes. Pressing ESC exits from either a submenu or the main menu without making any change.

Are your angles in radians or degrees?

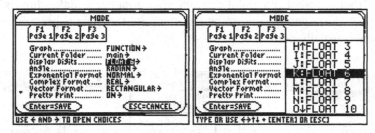

Figure 1.7 The MODE screen and the Display Digits submenu.

Since you are in calculus, you normally use the RADIAN setting. For situations where degrees are specified, you can change the MODE setting to DEGREE. However, it is recommended that you always leave the calculator in RADIAN mode and use 2nd_° to paste the degree symbol into the calculation where degrees are used as shown in Figure 1.8. The angle mode, either RAD or DEG, is shown on the status line.

Status line indicators

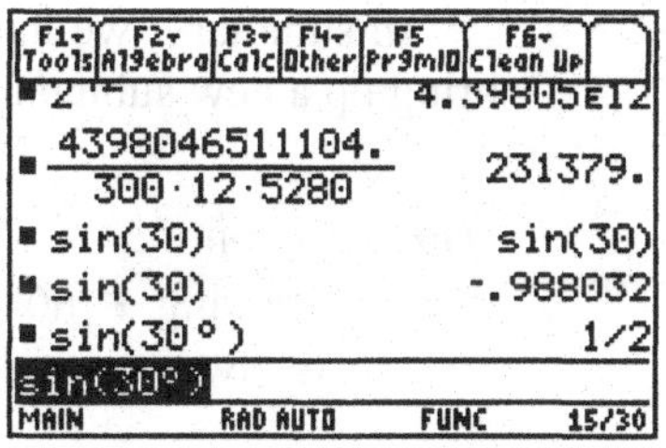

The status line on the screens we have seen show several MODE settings. We mention here those that we have seen so far. At the far left is the name MAIN: this is the current folder being used. The TI has the capacity of a small computer and folders help organize stored files. All our examples are in MAIN.

Figure 1.8 Using the degree symbol while in RADIAN mode.

The next indicator location to the right is usually blank; it shows whether a modifier key is in effect with one of the four symbols discussed above. Then the indicator RAD, shows that angles are measured in radians or else it shows DEG for degree mode. The AUTO indicator means calculations are displayed in exact form where possible unless there is a decimal in the input expression. The next indicator, FUNC, is the function graph mode; currently we are graphing functions. Finally you see the history count, telling you the number of stored history pairs and the current limit on history. Occasionally you may see a busy indicator in the far right when the TI is working on a long calculation.

> *Tip:*   If your output values are in an unexpected or undesirable format, check the MODE
> settings. If you are having trouble changing MODE settings, you may have forgotten to
> press ENTER the second time to save new settings.

A summary of menu use

There are several types of menus. The first is a pull-down menu from the tool bar at top of the
screen. In the first screen of Figure 1.9 you see the F2 (Algebra) menu gives quick access to
useful commands in algebra. Selecting one and pressing ENTER pastes the command on the
input line. The F2 (Algebra) menu has some special visual cues that enhance the menu
system. The down arrow beside 8 tells you that there are more menu items below. If you
arrow down to A:Complex, you see ▶ on the right which tells you there is a submenu. Press
the right arrow to see the submenu shown in the second frame. The third frame shows F1
(Tools); notice that the first entry is dotted which means that it is not available for use in this
context. If an ellipsis (...) symbol appears after a name, a dialog box is presented when you
press ENTER. Once you have become familiar with the menus, pressing the number (or letter)
of the item is usually preferable to using arrow keys.

Figure 1.9 Pull-down menus, submenus, and dialog box indicators.

In a dialog box, you either have a box to fill from the keyboard or a list to choose from
(indicated by →). To keep the settings you have made in a dialog box you must press ENTER to
save them. If you press ESC, they revert to the previous settings.

In summary, we have seen that menu choices either paste an item to the cursor location,
bring up a new submenu, or bring up a dialog screen.

> *Tip:*   A common mistake is to change settings and go directly to another menu (like
> pressing ♦_HOME) which does not save the changes to the settings. It is a good habit
> to always exit dialog boxes by using either ENTER or ESC.

Stored values and symbolic variables

In Figure 1.10, the store key STO▶, which appears on the screen as →, is used to save a
numeric value into a letter variable. Variable names can be up to eight characters in length,
but a name cannot start with a number. For example, if you wanted to repeatedly use the area
of a 10 inch radius pizza in calculations, you would enter:

$$\pi * 1 \varnothing \wedge 2 \text{ STO ALPHA\_a ENTER}$$

The variable name, a, can now be used in computations. In Figure 1.10, we find the difference in area of a 10 and 12 inch pizzas; the added two inches give almost 50% more. Then we calculate the cost per square inch of a 10 inch pizza that costs $12.99.

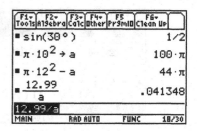

Figure 1.10 Using STO to store a variable.

A symbolic processing example

Although algebra is not a topic of this book, we show a simple example of using the TI for the algebraic manipulation of factoring. Many high school students suffer frustration from countless factoring worksheets, but the TI never complains — it just does it!

In Figure 1.11, three simple quadratic expressions with easy factoring are shown. Three different forms of factor are applied to the three quadratics equations. Each screen shows a stronger factoring capability. In the first frame with no specified variable, only the first quadratic expression factors. In the second screen a variable is named and now the second expression factors over the real numbers. In the third screen, using cFactor, all three factor with the third equation factoring over the complex numbers. When the TI cannot factor the expression with the specified constraints, its "answer" is the unsimplified quadratic.

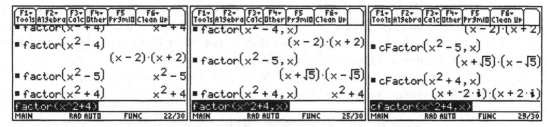

Figure 1.11 The factor (expression) uses integers while factor (expression, variable) uses all real numbers and cFactor (expression, variable) uses complex numbers (which includes the reals).

For this example you could use 2nd_ALPHA (A-LOCK) and type in the command factor. Or you could press F2 2 ENTER to paste the factor command. In general pasting is much faster. In either case, use simple edits on the input line to recycle the factor command as needed.

The TI answers are real numbers unless you specify that you want a complex answer. From this example you should be able to guess that in future examples pairs such as solve and cSolve, zeros and cZeros operate in the same manner; the leading c indicates complex answers are allowed.

Tip: The TI is not case sensitive, so typing cfactor instead of cFactor makes no difference. Regardless of how it is entered, the command appears in the history area in its standard form.

Defined and undefined variables

We end this chapter with a note about a frequent source of error in symbolic processing. Variables are either defined or undefined. A variable must be undefined when used symbolically. For example, in the previous factoring, if we use a as the variable in the expressions then a very different answer shows. See Figure 1.12. The factor command works numerically since $a = 100\pi$ from a previous use and thus gives a numeric answer in terms of π.

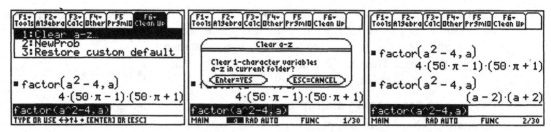

Figure 1.12 Defined variables work numerically. Undefined variables work symbolically.

There is no way to know if a variable is defined or undefined by its screen representation. If you want to check a variable's status, type its name and press ENTER: if it shows as a number, it is defined. An alternate, but less convenient, way of checking the status of a variable is with 2nd_VAR-LINK.

Cleaning up and clearing variables

The pull-down menu F6 (Clean Up) has two options for clearing variables. The first, 1:Clear a-z, was used in Figure 1.12. A dialog box is used to confirm this action. By using single letters for defined variables, you can quickly clear them all by using F6 before you start a new problem.

The second option, 2:NewProb, pastes NewProb on the entry line and you confirm this action when you press ENTER. So what is the difference? NewProb clears one-letter variables, but also clears the history area. It also does a few other things like deselecting functions which are discussed in the next few chapters.

The third option on F6 is a way of turning on and off custom menus. As noted earlier only default settings are used in this book.

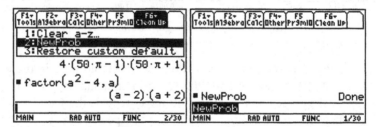

Figure 1.13 Use F6 (Clean Up) 2:NewProb to clear single letter variables and the history area.

Tip: Be sure the command NewProb is pasted on a clean entry line. If it is appended to any existing command on the entry line, an error message appears.

Tip: Longer named variables are deleted by using the command DelVar or through the 2nd_VAR-LINK menu.

CHAPTER TWO

DEFINING FUNCTIONS

The definition of functions and the use of functional notation are vital to success in calculus. In the next three chapters, we use the calculator's Graph/Table keys to define and evaluate functions, to make tables of values, and to graph functions. In short, we view functions analytically, numerically and graphically. In this chapter, we focus on the first key in the top row of Graph/Table keys: ♦_Y=.

Formula vs. function notation: using the Y= editor

Function notation is used in calculus, whereas formulas are used in algebra. So what is the difference and how are they related? They both express a relationship between variables. Let's take the famous formula for the area of a circle, $A = \pi r^2$. In precalculus you learned to write this in functional notation as $f(r) = \pi r^2$. The functional notation tells you <u>explicitly</u> which variable is the independent variable. Here it is r.

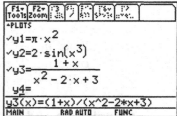

You can define up to ninety-nine functions in your calculator by using the function editor. The editing screen appears when you press ♦_Y=. The available functions are labeled y1, y2, ..., y99. To define a function, it is easiest to think of it in formula form. Let one of the y's be the dependent variable and make x the independent variable. For example, in our area formula we would define y1=πx^2, where r, the independent variable, is replaced by x.

Figure 2.1 Defining functions from the keypad in the Y= editor.

Functions are defined from the entry line using the keypad; see the examples in Figure 2.1. Notice that the entry line shows the keystrokes to input the function, but the definition area shows the pretty print form.

You can type a function definition directly onto the entry line, but some functions, such as the absolute value, can be pasted into a function definition from a menu. Press 2nd_MATH, 1:Number, 2:abs(ENTER to paste the notation in place at y4=, and finish by pressing x) as shown in Figure 2.2.

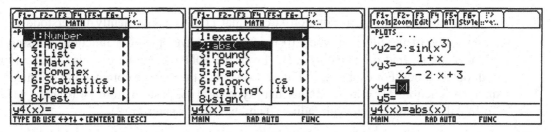

Figure 2.2 Pasting an expression from the 2nd_MATH menu to define abs(x).

Pasting from the CATALOG

Remembering the spelling and location of special symbols and functions within various menus can be tedious. This is why there is a built-in alphabetic listing of all functions and settings on the calculator. If you don't know the menus well, then the most convenient way to paste an expression into a function definition is to use the CATALOG key. In the catalog screen you move quickly to the function you want by pressing the letter key that starts its name. For example, if you want to use the secant (sec), then press CATALOG and you see the list beginning with the last entry you viewed in the catalog. Now press S (unless you are already there) and press ENTER. Luckily sec is the first entry. If the desired command was sum, it is faster to press T and scroll back.

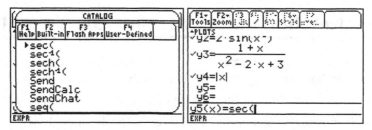

Figure 2.3 Defining a function using CATALOG. Pressing ENTER pastes the expression to the cursor position on the entry line.

Cleaning up and getting out: ESC or 2nd_QUIT or HOME

On the function editor screen, just as for the HOME screen, use the arrow keys to navigate. Use DEL and INS to edit. Pressing CLEAR deletes any definition. Use ESC to back out of menus or use HOME to directly return to the HOME screen.

Evaluating a function at a point

A benefit of the functional notation $f(x)$ is that $f(10)$ is conveniently understood to be the output value of the function when 10 is input. We have entered a function y1 using the Y= editor, and can find the value y1(10). From the HOME screen we can also define functions with other names. For example, in Figure 2.4 we use the Define command to define the same the area function, $f(r)=\pi r^2$. Rather than type Define, paste it from F4 (Other) 1:Define. To find the area of a circle with radius 10 cm, we simply enter f(10).

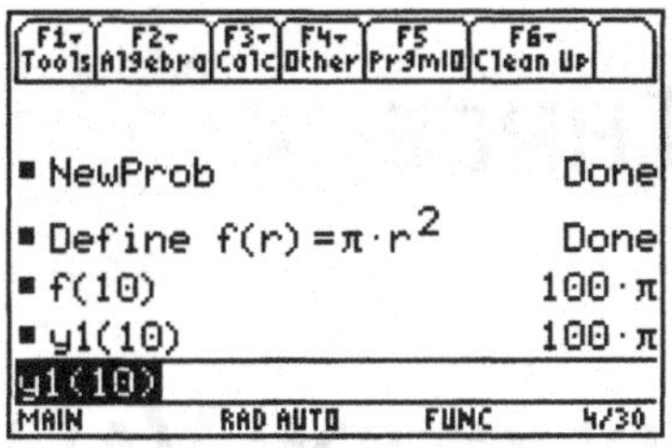

Figure 2.4 Evaluating f(10) on
the home screen.

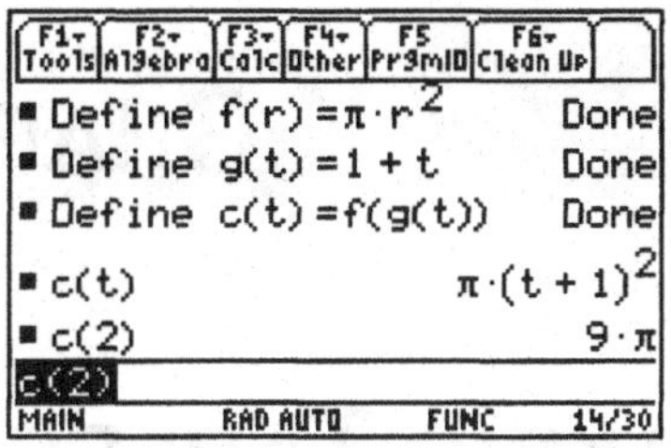

Figure 2.5 Creating a composite
function for area in terms of time.

New functions from old

In the next two examples, we create new functions from previously defined ones.

Composite functions: f(g(x))

Suppose an oil spill expands in a perfect circle and that the radius increases as a linear function of time. We can create a new composite function that expresses the area in terms of time. In Figure 2.5, define $f(r) = \pi r^2$ and $g(t) = 1 + t$, where t is in hours. Then define a new composite function c(t)=f(g(t)). Notice that c(t)=π·(t+1)² is the composite definition in terms of t. To find the area of an oil spill after 2 hours enter c(2); this gives the same answer as f(g(2)) (using ans(1) = g(2)).

A Malthusian example

In 1798, Thomas Malthus proposed that population growth was exponential and that food supply would grow at a linear rate. We return to the Y= editor where we model a food supply (per million persons) as y1 = 5 + .2x; this means that there is food for five million people in the base year and that each year afterwards the supply increases to provide for an additional 200,000 people. Press CLEAR as you enter these functions and the previous ones are erased. For the population (in millions), set y2 = 2(1.03)^x; this corresponds to an initial population of two million that increases annually by three percent. Let y3 = y1(x) - y2(x) and y4 = y1(x) / y2(x). These two new functions are a measure of excess food and a measure of food per capita, respectively. In Figure 2.6 these two measures are evaluated at 50 and 100 years from the base year. There is a shortage in the hundredth year (i.e., y3 is negative and y4 is less than 1).

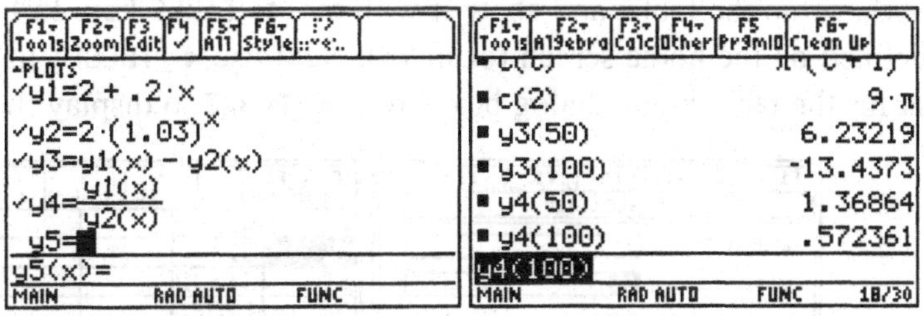

Figure 2.6 Food excess and food per capita in 50 and 100 years.

CHAPTER THREE

MAKING TABLES OF FUNCTION VALUES

This chapter's focus is making tables of function values. These values often reveal the nature of the functional relationship: are the values increasing? decreasing? periodic?

Lists of function values

To see a set of values for a function, you can evaluate a function with a list as the input variable which outputs a corresponding list of function values. For example, to find the area of circles with radii 10 cm, 50 cm, 100 cm and 200 cm. Define the area function and evaluate the function with a list as shown in Figure 3.1. Use the left/right arrow keys to scroll horizontally when data is too long to fit on the screen. Look for the symbols ◀ and ▶ on the right or left of the output line to indicate that there are more values off the screen in that direction. Notice in Figure 3.1 that the top menus have been changed by pressing ♦_Custom. This allowed the use of F2 ,6 to paste in Define f(x)=.

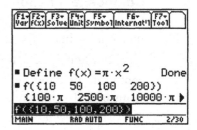

Figure 3.1 The values of the area function found as a set {10, 50, 100, 200}. Use arrows to scroll the answer set.

> **Tip:** If you used ♦_Custom and have the custom menu bar in effect, you can return to the regular menu bar by pressing ♦_Custom again; it is a toggle.

A table of values for a function

A more convenient way to see a list of values for a function, is to define one of the reserved Y= functions, setup the table, and show it with the ♦_TABLE key. In Figure 3.2, the function y1 was defined on the home screen (or in Y=). Next use ♦_TBLSET to set the beginning and increment for the table in the dialog box. Press ♦_TABLE to display the table; see Figure 3.2.

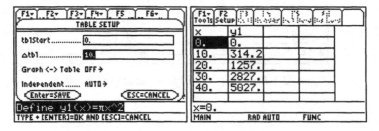

Figure 3.2 Making a table of values for the area function y1=πx².

The values beyond 40 are not shown on the first screen, but you can use the down arrow to see them. You can also use F2 (Setup) to return to the table setup dialog box and adjust the values. If there are other Y= functions evaluated in your table, disregard them for the moment, they can be turned off (deselected) as explained shortly.

Selected values for a table

Suppose you just want the function evaluated at some specific list of values, in table setup, you arrow down to the option Independent: Auto and use the right arrow to change from 1:AUTO to 2:ASK. Now press ENTER to save the change and press ♦_TABLE. The previous table values appear, but you can write over previous x-values and customize the list. See Figure 3.3. Some values are listed in scientific notation, but by highlighting the cell you see the full stored value.

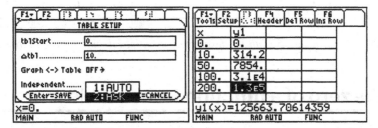

Figure 3.3 How to enter selected values rather than an incremented list.

Tip: You can use F1 9:FORMATS to adjust cell width.

Tip: If some of the rows already show values you want to include, use F5 (Del Row) to delete the ones you do not want and then add other values.

A table for selected functions

Recall the food supply and population functions from the Malthus model of the previous chapter. We want to enter these again but also want to keep our area function as y1, so we can y2 and y3 in the Y= editor; see Figure 3.4. Now press ♦_TBLSET to setup the table format from the dialog box and then display a table of function values using ♦_TABLE. Unwanted y1 values appear and we now show how to make them not appear in our table.

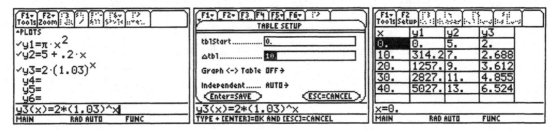

Figure 3.4 Using a table for selected functions.

Selecting and deselecting a function

When you want to see values for only certain functions, you can deselect the ones you do not want and select the ones you do want. In the Y= editor, move the cursor to function definition line and press F4 (✓). This is a toggle: if it was checked, it turns off; if it wasn't checked, it turns on. When you make a function definition, it is automatically turned on. To avoid having y1 displayed, deselect y1, as shown in Figure 3.5, so now only y2 and y3 show in the table display. Recall that F2 (Setup) is a convenient access to reset the table settings.

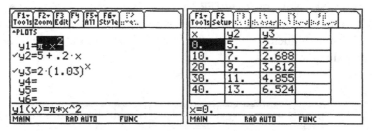

Figure 3.5 With y1 deselected, its values do not appear in the table.

Finding the zero of a function from a table

A natural question arises from the Malthus model: When is the food supply no longer sufficient for the population? Set y4=y2(x)-y3(x) to measure the excess food supply, and ask when the excess food function is zero. Since we know from previous evaluations that the excess is positive at year 50 and is negative by year 100, we set tblStart=50 and Δtbl=10 and see that the zero is between year 70 and 80. See Figure 3.6. Now use tblStart=75 and Δtbl=1. We could continue in this manner for a more precise value by setting tblStart=79 and Δtbl=0.1.

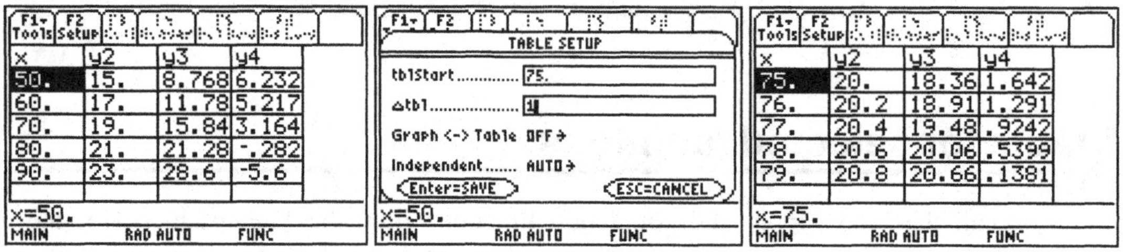

Figure 3.6 Searching for a zero of the food excess function.

Tip: Deselected functions are still active for calculations when used in other function definitions.

Tip: Change a Y= function definition from inside a table by using F4 (Header).

CHAPTER FOUR

GRAPHING FUNCTIONS

This chapter completes our investigation of the Graph/Table keys and shows how to graph the functions that we have defined.

Basic graphing

Graphing is like the 1-2-3 of taking a picture with a camera.

- *Select your subject(s)*. To select a function, recall that you use F4 (✓) from the Y= editor. (Deselect or clear functions that you do not want to graph.)
- *Frame them properly*. Press ♦_WINDOW and set the x- and y-window boundaries.
- *Click to take the picture*. Press ♦_GRAPH.

The hard part of photography is getting the subject both in the picture and looking good. The WINDOW menu controls the picture frame. Figure 4.1 shows the settings and graph for a picture of the function y1=πx². The function's graph uses too little of the screen. The last frame shows an improvement made by changing two settings in the WINDOW screen. To save space a split screen mode is used in Figure 4.1. This feature is discussed in Chapter 5 and here it is recommended to uses a full screen

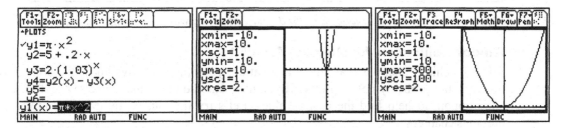

Figure 4.1 The basic graph sequence: ♦_Y=, ♦_WINDOW, ♦_GRAPH and then an improvement by changing the WINDOW settings. Split screen mode is used in this display, Use full screens.

Window settings: xmin, xmax, xscl, ymin, ymax, yscl, xres

The window setting variables are as follows:

xmin sets the left edge of the window as measured on the horizontal axis,

xmax sets the right edge of the window as measured on the horizontal axis,

xscl (x-scale) sets the width between tick marks on the horizontal axis,

ymin sets the bottom edge of the window as measured on the vertical axis,

ymax sets the top edge of the window as measured on the vertical axis,

yscl sets the width between tick marks on the vertical axis, and

xres (x-resolution) sets the selection density of values to plot (2 is the default setting).

Finding a good window

Like photography, the setting of the window is an art. It is rare to know the ideal window before graphing; trial-and-error experimentation is usually required. Use the F2 (Zoom) menu to start the process. Zoom is available on top of the Y=, WINDOW and GRAPH screens. With thirteen choices (including a submenu), it is hard to remember the whole menu. Let's begin with three items that address a common need — a quick window setting.

Special settings: ZoomStd, ZoomDec, ZoomTrig

There are three Zoom menu options that automatically set the window to special settings and graph the selected functions, all in one keystroke. These special settings are shown in Figure 4.2. They are especially helpful when you are graphing common functions (such as polynomial, exponential, and trigonometric) with graphs close to the origin. The 6:ZoomStd (Zoom Standard) setting often works well as a good first view. The 4:ZoomDec (Zoom Decimal) option gives what is called a *nice* window because the *x*-values used are from -7.9 to 7.9 by tenths. The nicety of this is explained in the tracing section just ahead. For trigonometric functions, the obvious first choice for graphing is 7:ZoomTrig (Zoom Trigonometry); it uses *x*-values from -(79/24)π to (79/24)π (shown in decimal form; think of it as just over -3π to 3π). Using ZoomStd resets xres to 2, but the other two do not change the xres setting from its current value.

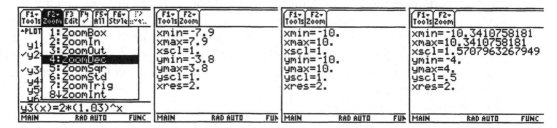

Figure 4.2 F2 (Zoom): ZoomDec, ZoomStd, and ZoomTrig window settings, respectively.

Window adjustment for Malthus: ZoomFit

When graphing functions that model a situation, you almost always know the domain of the function, but probably not the range. If you have entered the domain, use A:ZoomFit, another Zoom option, to help find the best *y*-values. Let's try this out on the Malthus model of the previous chapters.

Malthus never published a graph; he used only numerical and analytical expositions. Some of his readers didn't see his concern; he needed a graph. Enter or turn on the two Malthus equations. Start with ZoomStd to see some graph but because we want to look from the base year to a century in the future, we use ♦_WINDOW and set xmin=0 and xmax=100. Now the range of the functions is difficult to guess, so we leave the setting (i.e., ymin=-10 and ymax=10) and use ZoomFit. See Figure 4.3.

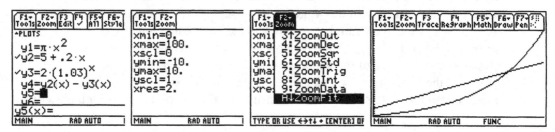

Figure 4.3 Using Y=, WINDOW, and ZoomFit to get a graph with a good range.

Tip: Set xscl and yscl to zero as you adjust a window, then choose helpful values when the final window size is found.

Tip: Since there are no numeric labels on the graph, it is often necessary to use the ♦_GRAPH and ♦_WINDOW keys to check on the window dimensions. This problem of knowing where you are on an unmarked graph is the next topic.

Identifying points on the screen

By pressing any arrow key when a graph screen is showing, a cross-hair cursor appears in the center of the screen and the *x*- and *y*-values of the cursor point are displayed. See Figure 4.4. This cursor is called the free-moving cursor. You can use the arrow keys to navigate this cursor to any point on the screen. By placing the free-moving cursor on the graph of a function, you can display the approximate coordinate values of the point $(x, f(x))$. The *x*- and *y*-values are labeled xc and yc. If you press ENTER or CLEAR, the free-moving cursor mode is canceled.

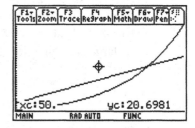

Figure 4.4 The free-moving cursor, activated by an arrow key.

Reading function values from the graph: Trace

The values of the function are identified in Trace mode. Pressing F3 (Trace), from the graph screen activates the trace cursor. Figure 4.5 shows the trace mode displaying the *x*- and *y*-values of the current trace cursor location and in the upper right corner the number of the graph being traced. Use the right/left arrows to move along the graph of the function.

You can also enter a specific *x*-value as shown in Figure 4.6. In Trace mode, just type the desired value and press ENTER. (Pressing F3 again moves to the nearest point normally displayed on the graph.) In the last panel of Figure 4.6, the trace cursor is switched to the other graph with the up and down arrows. In this case, we started on y2 and went to y3. Pressing CLEAR cancels the Trace mode.

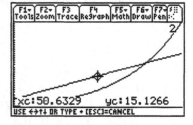

Figure 4.5 The Trace cursor on a function.

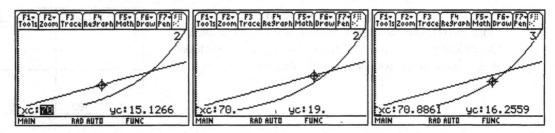

Figure 4.6 Right/left arrows move along the graph. Up/down arrows move between selected graphs.

Making better trace values: ZoomInt

As you arrow right (or left) using Trace, you see x-values with long decimal expansions. There is little need to know the food supply after exactly 70.8861 years. For a rough idea about values, we can ignore the extra decimal digits. However, using F2, 8:ZoomInt resets the window so that the trace values are integers. You are first asked for the desired center of the graph and we move the cursor to xc=50 and yc=20.2186 before pressing ENTER. Now use Trace to see that the x-values are all integers. Use ♦_WINDOW to see the adjusted settings.

How to make a nice window (an optional adventure)

The screen is 159 pixels wide and this is why the ZoomDec setting is so nice: with xmin=-7.9 and xmax=7.9, the resulting x-values are 79 negative tenths, zero, and 79 positive tenths, a total of 159 pixels. With ZoomInt in the last example, the x-values started at xmin=-29 and ended at xmax=129, 159 values altogether (counting zero). The same holds for ZoomTrig which spans 159 intervals of length π/24. For the Malthus graph, a nice window starts at xmin=0, ends at xmax=158. In general, for a graph starting at xmin=0, set xmax=15.8*n, where n is large enough to let 15.8·n span the x-values you want to include.

Tip: The xres setting affects the tracing. A lower xres allows more trace points.

Tip: Don't confuse 8:ZoomInt with 2:ZoomIn.

Panning a window

There is an old story about blindfolded people describing an elephant from different perspectives; their guesses included wall, tree, and snake. Sometimes functions are like elephants: you need to take the blindfold off to see the whole picture. In case the subject is too big to fit in one window, we move the window frame to see what is to the left or the right or above or below the current view. This is called panning. Let's take an example of a logistic equation and start as if we knew nothing about it.

$$f(x) = \frac{1000}{1 + 9e^{-0.05x}}.$$

Enter the function: y1=1000/(1+9e^(-.05x)). Get a first look by using ZOOM 6:ZStandard. We see no graph. This is the -10 ≤ x ≤ 10 and -10 ≤ y ≤ 10 window and no function points appear in that window.

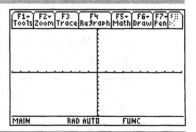

Press F3 Trace to find some x, y value. We see yc: is close to 100. We can reset the window by hand or use a Quick Zoom technique. Whenever you are in trace mode, pressing ENTER shows some part of the graph.

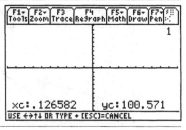

Press ENTER to Quick Zoom. This pans the window up and places the trace cursor at the center. Quick Zoom always brings some part of the graph into the window. After ENTER press F3 (Trace) to see where you are.

At this point you can use ZoomIn or ZoomOut to see more of the graph.

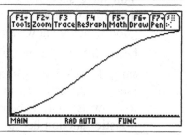

Instead, let's try setting the *x*-values over a broader interval, say $0 \leq x \leq 100$, and then using ZFit to see a more complete graph.

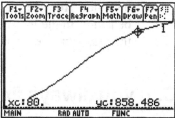

To explore at a specific point, like $x = 80$, press Trace 8 0 ENTER which evaluates the function and displays the trace cursor there. The entered value in trace mode must be between xmin and xmax. This technique cannot be used to move (pan) the current window.

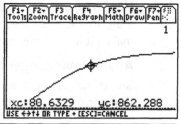

As before, pressing ENTER for Quick Zoom pans to a graph centered at $x = 80$. To investigate values further to the right, and outside the current window, hold down the right arrow key.

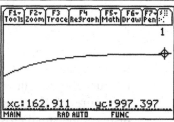

When the trace cursor reaches the right side of the screen, the screen pans to the right. Pressing 2nd_ right or left arrow puts the trace cursor in *turbo* mode — the cursor moves in larger steps.

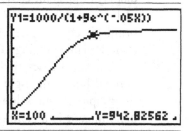

After seeing various aspects of the function, press WINDOW, enter good window values to see the all the important behavior of the graph.

In summary, we used two techniques to move a window without giving specific numbers. The Quick Zoom command is Trace ENTER, which pans and centers the window on the trace cursor. Secondly, the trace cursor, when moved to the left of the Xmin value or to the right of the Xmax value, pans the window horizontally.

> *Tip:*   The free-moving cursor does not pan the screen.

What is a good window?

You have seen the trial-and-error approach to finding a good window, but this begs the question: what is a good window? For a function serving as a model of a physical situation, a good window shows the function's graph for the relevant domain. For example, there would be no interest in negative values of the area function, $A = f(x) = \pi x^2$. But considering the same function as a purely algebraic quadratic function, you would choose a window that includes negative x-values to see all of its important behavior. Whenever possible, you want to show asymptotic behavior. For example, we needed to see past $x = 100$ in the graph of the logistic function because that function is approaching the line $y = 1000$. We call this the end behavior. However, if we concentrate solely on the end behavior, we might blur some local behavior. In the logistic example, an important local behavior is the point of inflection where the graph changes from concave up to concave down. (Concavity is discussed in Chapter 10.)

Can you always find a good window?

No. There are pathological functions that we cannot graph and others that require than one view to show all their important behavior. This case occurs when the end behavior view makes it impossible to see the local behavior and vice versa.

Asymptotic dangers: Beware of graphs with vertical lines

Sometimes the graphing calculator leads you astray. The most common case is rational functions. Let's take the blind graph approach to

$$p(x) = \frac{x^2 + 2x + 30}{x - 4}$$

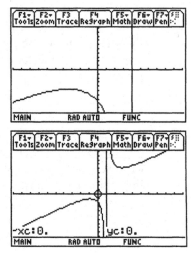

First, we enter y1=(x^2+2x+30)/(x-4) (don't forget the parentheses) and use ZoomStd to get some idea about the function. See the top frame in Figure 4.7. We need to see more of the graph. We now use 3:ZoomOut. This displays the small cross-hair cursor to designate the center of the expanded window to come. The current center (the origin) is OK, so press ENTER. The result is a window with both domain and range four times as big.

In both frames of Figure 4.7 we see an unexplained vertical line to the right of the origin. Could this be part of the graph? A quick look at the function tells you that the denominator is undefined at $x = 4$. Recall from precalculus that this line is called an asymptote. But the calculator did not draw it as an asymptote. The calculator draws graphs

Figure 4.7 Graphing a rational function.

by connecting special x-values that are found by starting at xmin and adding increments of (xmax - xmin) / 159. In this case, to the left of $x = 4$, $f(3.5443) = -108.956$, and to the right of $x = 4$, $f(4.55696) = 107.512$. (You can verify these values using Trace.) Connecting these values gave the vertical line.

At the origin of the bottom screen a circle is visible. This is left by ZoomOut to be in the ready state for the next ZoomOut, just press ENTER for more zooming.

> **Tip:** One way to prevent connection across an undefined point is to create window settings with the undefined value of *x* being an evaluated point between xmin and xmax. For example, the line disappears using xmin=-32 and xmax=40. The trace cursor 'shows' that the value is undefined at *x* = 4 by not showing a *y*-value there.

Changing plot style

An easier way to remedy the connection problem is to change the style mode in the Y= editor from 1:Line to 2:Dot in the F6 (Style) menu as shown in Figure 4.8.
The other styles are:

    3: square dots
    4: bold connected line
    5: round cursor without path
    6: round cursor with path
    7: shading above
    8: shading below

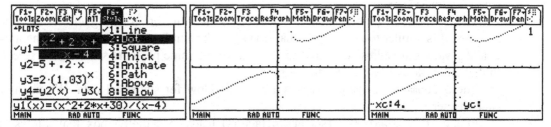

Figure 4.8 Changing the graph format to Dot display. Tracing an undefined value at x=4.

An inaccurate graph

We found the calculator may connect dots when it should not. A similar distortion occurs when the resolution is insufficient to display a function. Consider

$$f(x) = \sin\left(\frac{1}{x}\right)$$

There is no way this function can be accurately graphed if the origin is shown. Several attempts are shown in Figure 4.9. The xc:0. and yc:0. displayed in the second screen is misleading; it is not a trace value, ZoomIn leaves a prompt for center. The third screen is in Trace mode and correctly indicates that *y* has no values when *x* = 0.

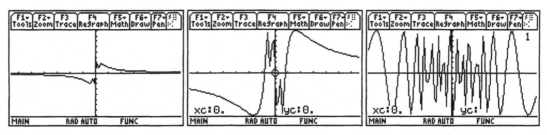

Figure 4.9 Three attempts to graph y1 = sin(1/x) accurately using the window settings ZoomTrig, ZoomIn, and -.1 ≤ x ≤ .1.

> **Tip:** In writing mathematics, it is good style to write decimal values less than one with a leading zero. Therefore, in the caption of Figure 4.9, we should have written $-0.1 \leq x \leq 0.1$. However, when Ø.1 is entered in the calculator, it is converted and shown as .1. For this reason, we often break with style and not use a leading zero.

Other Zoom options

For the sake of completeness, we mention the other choices on the Zoom menu. The 1:ZoomBox option is similar to ZoomIn and ZoomOut in that it displays the small cross hair cursor on the graph. Move this cursor with the arrow keys to the screen location where you want a new window to have a corner, press ENTER, then move the cursor to the diagonal corner desired (you see the rectangle on the screen as you move the cursor), and press ENTER again. The new window is the rectangle you defined.

The B:Memory submenu includes the 1:ZoomPrev option which returns you to the previous settings. This is handy if you have changed the window settings and the graph is worse. You can also keep one setting in memory: store a window setting with 2:ZoomSto and recall it with 3:ZoomRcl.

The 5:ZoomSqr (zoom square) selection is helpful for graphing circles or graphing inverses. It is similar to ZoomFit in that it is a one-keystroke grapher that fixes one axis. The x- and y-values are adjusted so that the axes have the same physical scale on the screen (i.e., so that circles look like circles, not ellipses).

The 9:ZoomData selection graphs a good window for statistical data. We use this in a later chapter.

Finally, C:SetFactors... (set zoom factors) leads to a dialog box that allows you to change the setting for multiplier/dividers xFact, yFact and zFact used in ZoomIn and ZoomOut (zFact is used in three-dimensional graphing.) The default setting of 4 is good, but sometimes 2 is more convenient.

CHAPTER FIVE

CALCULATING FROM A GRAPH

We have seen how to set up and graph a function. In additional, we have used F2 (Zoom) and F3 (Trace) from the graph screen. In this chapter we see how to use other features of the pull-down menus from the graph screen.

An important menu on the graph screen is F5 (Math). In later chapters it is used frequently to access the calculus items on this menu. In this chapter we restrict our attention to the Math menu items used to identify special points on a graph. The chapter concludes with instructions on split screen graphing and graphing inverse functions.

Finding roots: Zero

We first use the F5 Math, 2:Zero to find a zero (root) of a function. We return to the Malthus model (y2 and y3 as shown below) and consider the food excess function y1=y2(x)-y3(x). The zero of this function has an important meaning: it is when we start having a food shortage.

Although not essential, we first turn off (deselect) y2 and y3. We use the ZoomStd window to start and adjust it by making xmin=0, xmax=100, and xscl=10 in the WINDOW screen. Press ♦_GRAPH to see the food excess graph. It has a zero just before $x = 80$.

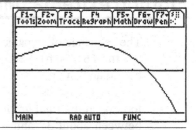

Press F5 (Math), 2:Zero. The F5 pull-down menu contains the most commonly used items in calculus. The F5 Math menu on a graph screen should not be confused with 2nd_MATH menu (above the 5 key).

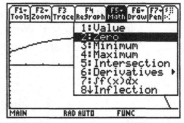

We are prompted for a lower bound. Since we are to the left of our zero, we just press ENTER. Optionally, you can enter an x-value.

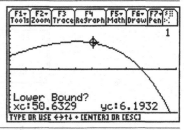

After pressing ENTER, a right facing arrow on the screen marks the lower bound and you are prompted to enter an upper bound. We enter 90 or arrow to the right to an *x*-value large enough that the function values are negative.

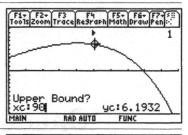

Press ENTER and the root is shown graphically and a numerical value is shown at the bottom of the screen. In some cases the yc value is not exactly zero but has an exponent (E-12) making the *y*-value close enough to zero for all practical purposes.

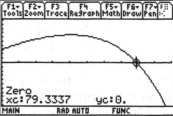

Tip: The special points of the F5 (Math) menu are only found between the current xmin and xmax settings.

Tip: The closer the bounds, the faster the TI finds the zero.

Finding extrema: Minimum and Maximum

The sequence of steps is the same whether you are finding a zero, a maximum, or a minimum. On the F5 (Math) menu, both 3:Minimum and 4:Maximum use the same request for a lower bound and an upper bound.

For the excess food function, we might ask in what year the excess is a maximum. By looking at the final graph in the above sequence, we see the graph is a horizontal line segment between about 37 and 45. This is misleading because the values of the function are not a constant on this interval, as tracing would show. Rather, the resolution of the calculator screen is limited. We use 4:Maximun to find the highest value. In Figure 5.1 we enter a lower bound of 37 and an upper bound of 45 as numeric estimates and find the maximum at just over 41 years.

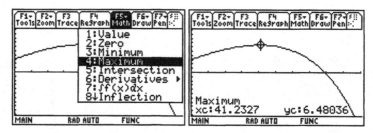

Figure 5.1 Finding the maximum of a function within the interval: Lower Bound=37, Upper Bound=45. (Bound selection screens are not shown.)

Finding an intersection of two graphs

A typical task is to find where two functions are equal. Analytically, this means finding the zero of the difference function as we have done. Graphically, this means finding where the graphs of the two functions intersect.

We reset the window for $0 \leq y \leq 40$ and select y2 and y3, the food and population functions. From the F5 (Math) menu, select 5:Intersection. You are asked to identify the first curve; press ENTER to accept the default choice (identified by the flashing trace cursor). Select the second curve by pressing ENTER. If more than two curves are graphed, choose by using the up or down arrows. The calculator now prompts you for a lower bound and an upper bound; these must enclose the desired intersection point. (This is necessary since some graphs have multiple points of intersection.) We see that the intersection point is at 79.3337 years. We got the same answer as we did when finding the zero of the difference function — we must have done it right!

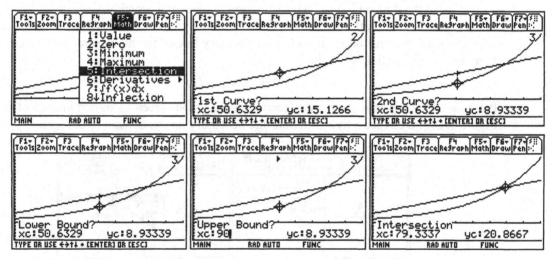

Figure 5.2 Finding the intersection of two functions within an interval.

Split screen graphing

The TI screen is quite wide in comparison to its height. At times it is helpful to see two screens at the same time, say a graph and its window values. The split screen option is set from page 2 of the MODE menu. There are three different settings for Split Screen:

1:FULL
2:TOP-BOTTOM
3:LEFT-RIGHT

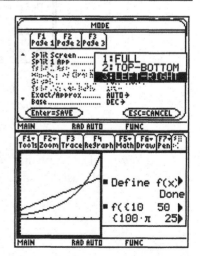

Figure 5.3 Split Screen options in MODE (Page 2).

We only use 1 and 3. Move the cursor to 3 and press ENTER twice. The lower screen in Figure 5.3 has a graph and a home screen. To move from one half-screen to the other, press 2nd_APPS (the icon is a split screen with arrows).

The easiest way to exit the split-screen mode is to press 2nd_QUIT from the entry line. This means that if you are not on the entry line, you need to use 2nd_QUIT twice in succession. You can also use MODE and change the setting.

Although used in this book for exposition, the split screen mode is not recommended for general use.

> **Tip:** Notice that the same graph in a split screen has a different ratio for the *x*-values. Some Zoom options set a different window for a split screen.

Graphing inverse functions

Recall that for any function *f*, the graph of f^{-1} is the reflection of the graph of *f* about the line *y* = *x*. We use the split screen mode to graph an inverse function since half a screen is nearly a square graph and best shows the relationship.

In Figure 5.4, we graph y1=cos(x) and its inverse, y2=cos⁻¹(x), in a ZoomDec window. Some of the graph's reflection across the *y* = *x* line is missing. The reflection of cosine is not a function so the domain of cos⁻¹(x) is restricted to make it a function.

To graph a full reflection of the function, we use the F6 (Draw) pull-down menu and select 3:DrawInv. This pastes DrawInv on the entry line of the home screen and we paste the function so that the full command is

DrawInv cos(x)

Pressing ENTER replaces the home screen with the graph screen as shown in the last frame of Figure 5.4. You can see why restrictions are necessary on the domain of cos⁻¹(x).

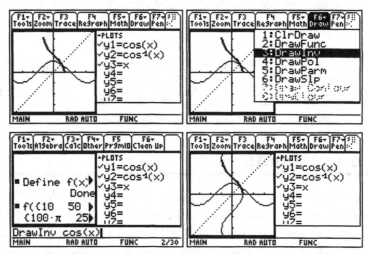

Figure 5.4 A function and its inverse, where the inverse is not a complete reflection across the line y = x. DrawInv gives a complete reflection.

> **Tip:** The graph drawn by DrawInv cannot be traced nor can any of the F5 (Math) options be used for it.

CHAPTER SIX

SOLVING EQUATIONS

This is our first use of the computer algebra system. The calculator works symbolically with variables, giving both symbolic and numeric answers. This power brings with it a requirement for careful syntax. Learn the commands carefully; it's worth the effort.

Solving a quadratic equations

In the last chapters we showed how to graph functions to find their zeros. A more direct approach without graphics is available in the F2 (Algebra) drop-down menu above the home screen. The F2 (Algebra) menu, offers two ways, 1:solve or 4:zeros, to find zeros. In Figure 6.1 we see a comparison of these two commands. To solve a quadratic equation you need two things: the equation (or expression) you want to solve and the name of the independent variable.

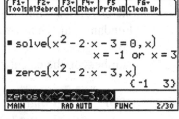

Figure 6.1 Using solve *and* zeros *with a quadratic expression.*

The command solve allows a general equations which need not have zero on one side; the zero command requires an expression that it sets to zero while solving. For many purposes, the output of zeros is in a more convenient form. For example, the set of zeros can be the input (as a list) for a function and evaluated. The advantage of solve is that it allows input of an equation in its original form, which is safer, since no intermediate algebra is needed to transform the equation to an expression.

The calculator can perform algebra with symbolic parameters. So, for instance, it can give us the famous quadratic formula. Our first effort, shown in Figure 6.2, just spits back our input equation with a strange number 39.9424. There are two problems: we did not clear all the variables (c had a previous value of 39.9424) and we entered the equation incorrectly. Look carefully to see that there is no multiplication dot in the history area between a and x^2. We entered ax^2 and the calculator understood that to be the two-letter variable ax squared. Despite our intentions, we entered an equation in three variables, ax, bx and c, and asked it to solve in terms of another variable, x, not present in the equation. No wonder the TI couldn't do anything! Using F6 (Clean Up) and re-entering the equation with $a*x^2$ and $b*x$ gives much better results.

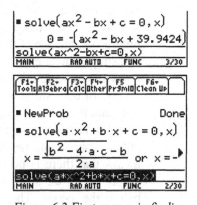

Figure 6.2 First, errors in finding the quadratic formula. Secondly, a success.

> *Tip:*   Some longer variable names are reserved words and must be avoided in symbolic expressions. These are listed in the *TI Guidebook*.

> *Tip:*   Any time you enter an expression or equation in a command, you must identify the independent variable. The Too few arguments error message is the reminder.

Solving a quadratic equation with complex solutions: cSolve

If solve finds the equation is an identity, it responses with true, i.e., it is true for all reals.

When we try to solve the quadratic equation $x^2 + 5 = 0$ in Figure 6.3, the result is a terse false. This means that the equation has no real solutions. The F2 (Algebra) menu has a submenu A:Complex with complex number commands. Using cSolve gives the complex number solutions of the quadratic equation. Similarily, cFactor works using complex numbers, not just the reals, and gives the factorizations into linear terms guaranteed by the Fundamental Theorem of Algebra. The appendix of this book gives more information about using complex numbers on the calculator.

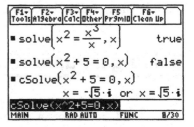

Figure 6.3 Complex analogs of solve *and* factor *from the F2 submenu* A:Complex.

> *Tip:*   If you find solve or factor is inadequate and you want to use cSolve or cFactor, just add c in front of the command. This avoids a trip to the complex submenu. Also it is not necessary to capitalize S and F on the input line.

Solving non-polynomial equations

High school algebra is sometime taught as though it can be used to solve any kind of mathematical problem. Yet there are many equations, even some polynomials (of fifth and higher degree), that have no algebraic means of solution. But they do have numerical solutions. For example, consider the four equations in Figure 6.4, each sets the basic exponential function equal to a simple linear expression.

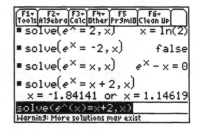

The symbolic solution to $e^x = 2$ is derived by taking the inverse of both sides: it is an exact answer. (Be sure to use ♦_e^x, not the letter e when entering the equation.) Next we look for the point where the exponential function equals -2, but of course this never occurs and the response false means that there is no real solution. This agrees with the graph of $y = e^x$; since the exponential curve never falls below the x-axis. The calculator's response to the third

Figure 6.4 Several types of output from solve.

solve command should be false since the graph of $y = e^x$ does not intersect $y = x$. However, this is an instance of the TI doing what it can and reporting back no conclusion. The last case gives two numeric answers. Realize that these are approximations; there is no way to use an inverse and get an exact solution as we did in the first case.

Equations with an infinite number of solutions

How does `solve` handle answers for periodic functions, which may have infinitely many answers? We consider a simple case, sin(*x*)=1/2. In the first entry of Figure 6.5, we use .5 for 1/2 which results in a decimal answer. (When the Exact/Approx mode is set to AUTO, a decimal in the input causes numeric answers to be decimal approximations.) Repeating the command with the fraction 1/2 gives the answer in exact form. The symbols @n2 and @n3 signify any integer. The second answer is often written as $5\pi/6 + 2n\pi$. Until we clear variables, successive answers of this sort have the symbols @n3 , @n4, etc.

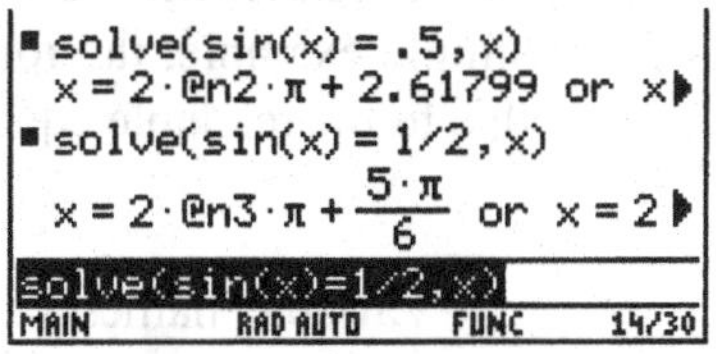

Figure 6.5 Solve answers when there are infinitely many solutions.

> *Tip:* The infinite answer variables @n# are most easily reset by using the F6 (Clean Up) option 2:NewProb.

Conditional solutions

A powerful feature of the TI is the ability to impose conditions on solutions. It is rather simple but often the command is so long that it requires scrolling to see it all on the input line. For example, suppose we want only positive answers. In Figure 6.6, at the end of the first `solve`, command we add the vertical bar symbol "|"and the condition x>0. Another use of this technique is that you can enter an equation with letter parameters and define any known parameters as a condition. These definitions are not permanent, but only used for that entry. More than one condition requires the conjunction "and" which is conveniently entered using the CATALOG. The complete entry for the second screen is:

$$\text{solve(a*x^2+b*x+c=0,x)|a=1 and c=2}$$

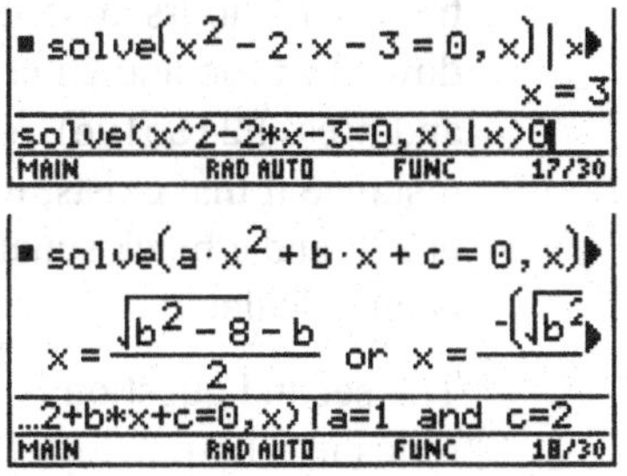

Figure 6.6 Using solve with conditions.

Scrolling is required to see all of the input and all of the solution.

Simultaneous solutions to systems of equations

The `solve` command works with simultaneous solutions to systems of equations. In Figure 6.7 we find the intersection points of the unit circle, with its equation, $x^2 + y^2 = 1$, and the parabola, $y = x^2$. Use and to join them and use {x,y} to specify a two variable solution. The complete entry is:

$$\text{solve(x2+y2=1 and y=x2, {x,y})}$$

The whole solution requires scrolling to see the *y* value and to see that there are two pairs of solutions.

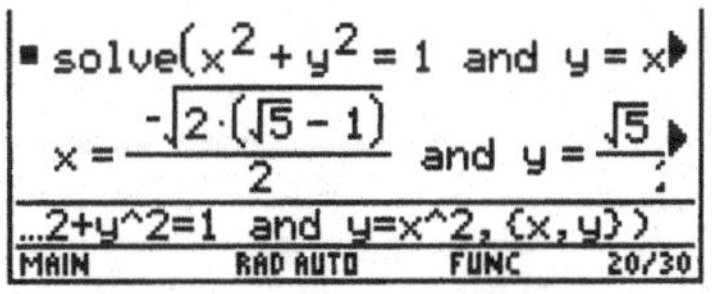

Figure 6.7 Using solve with a simultaneous systems of equations.

Using a numerical solver: What if?

For more versatile solving the Numeric Solver in the APPS menu is handier than solve. For a simple example, the formula for motion is "distance equals rate times time" and we translate this to the readable equation:

distance = rate * time

Variable names can be from one to eight letters long, so we choose full descriptive names. Press APPS, Numeric Solver and enter the new equation as shown below. We then play "what if" by setting any two of the equation's variables and using F2 (Solve) to calculate the value of the remaining variable.

Tip: Before you press F2 (Solve), check to see that the cursor is on the line of the variable for which you want a solution.

When you enter the Numeric Solver, any previously used equation is still there. Use CLEAR to start a fresh equation. F5 (Eqns) gives a menu of known equations. After entering your equation, press ENTER. A list of your variables and their current values appears — if they are not blank, variables may have been previously defined.

```
F1- F2  F3-   F4      F5   F6
Tools Solve Graph Get Cursor Eqns Clr a-z...
distance=rate*time
  distance=|
  rate=
  time=
  bound={-1.E14,1.E14}

MAIN        RAD AUTO      FUNC
```

The first question we answer is what distance do we travel in 4 hours at 55 mph. Enter 55 for rate, arrow down to time and enter 4, arrow back up to distance and press F2 (Solve). You need not clear the value of distance if that exists; it is treated as a guess. The answer is 220 and a black square on the left certifies that it has been calculated.

```
F1- F2  F3-   F4      F5   F6
Tools Solve Graph Get Cursor Eqns Clr a-z...
distance=rate*time
■distance=220.
  rate=55.
  time=4.
  bound={-1.E14,1.E14}
■left-rt=0.

MAIN        RAD AUTO      FUNC
```

The second question is what is our rate if we travel 300 miles in 5 hours? Enter 300 for distance and notice the black square disappears, arrow down to time and enter 5, arrow back up to rate and press F2 (Solve). As expected the answer is 60 and certified by the black square.

```
F1- F2  F3-   F4      F5   F6
Tools Solve Graph Get Cursor Eqns Clr a-z...
distance=rate*time
  distance=300.
■rate=60.
  time=5.
  bound={-1.E14,1.E14}
■left-rt=0.

MAIN        RAD AUTO      FUNC
```

The last question is what is the time to travel 500 miles in at a rate of 55mph? Enter 500 for distance (black square disappears as soon as you change distance), arrow down to rate and enter 55, arrow back up to time and press F2 (Solve). As expected the answer is about 9 hours.

```
F1- F2  F3-   F4      F5   F6
Tools Solve Graph Get Cursor Eqns Clr a-z...
distance=rate*time
  distance=500.
  rate=55.
■time=9.09090909091
  bound={-1.E14,1.E14}
■left-rt=0.

MAIN        RAD AUTO      FUNC
```

Tip: The TI-89 keyboard makes it awkward to type long names for variables, using A-LOCK helps, but you may want to use a more simply written equation like d=r*t.

CHAPTER SEVEN

THE LIMIT CONCEPT

A fundamental difference between precalculus and calculus is the application of the limit. In precalculus, we define average velocity over some time period of positive length. For example, if you drive 200 miles in four hours, then you averaged 200/4 = 50 miles per hour. But looking at the car's speedometer, you see the speed at a given moment, like 57, this is the instantaneous velocity. In this chapter we use tables and graphs to investigate the idea of the limit and then introduce the limit command.

Creating data lists and finding average velocity

The following table shows heights, h, of a grapefruit thrown in the air after t seconds. Find the average velocity over periods of one second. (This is quite simple, but it allows us to cover several general aspects of lists on the calculator.)

t (seconds)	0	1	2	3	4	5	6
h (feet)	6	90	142	162	150	106	30

Table 7.1 Grapefruit height, per second.

First we put the data into two lists named t and h using the set bracket notation

$$\{\emptyset,1,2,3,4,5,6\} \rightarrow t$$

$$\{6,9\emptyset,142,162,15\emptyset,1\emptyset6,3\emptyset\} \rightarrow h$$

A comma is the most convenient separator but it is transformed to a space in the history area. To know what the first average velocity value should be we calculate it by hand: (90 – 6) / (1 – 0) = 84. In Figure 7.1, we use ∆list to do the repetitive calculations and display the whole list at once.

The ∆list is found in the CATALOG; it subtracts successive entries in a list and outputs a list that has one fewer entry. This example shows that lists can be used for arithmetic calculations and the operations are done term by term. It is noted that in this example the time intervals are all 1 and the division by ∆list(t) is unnecessary.

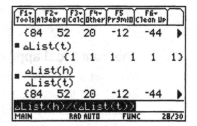

Figure 7.1 Finding the average velocity from data.

What does the limit mean graphically and numerically?

The concept of limit is the foundation of calculus. For example, the key to calculating an instantaneous velocity is to let the time period become closer and closer to zero. Using data lists, this is virtually impossible. If we know an expression for a function, say displacement $s(t)$, then the average velocity from time a to time $a + t$ can be written

$$\frac{\Delta s}{\Delta t} = \frac{s(a + t) - s(a)}{t}$$

Notice that the time period t is in the denominator of the average velocity, so letting it reach zero would mean dividing by zero — certainly an error. But limits avoid that problem: the idea is to see if the function approaches some value L as the t-values get closer to zero, without ever reaching it.

The graphical approach to limits

Let's see how the limit works with a particular function. A function g is defined as an average velocity function with $s(t) = t^3$, $a = 1$ and $t = x$.

$$g(x) = \frac{(1 + x)^3 - (1)^3}{x} = \frac{(1 + x)^3 - 1}{x}$$

In Figure 7.2 the graph of $g(x)$ appears to have a value of 3 at $x = 0$. However, when we use trace there is no y-value shown when $x = 0$. For x-values on each side of zero the trace cursor appears and it doesn't take a rocket scientist to figure out that the y-values are getting close to 3 as x gets close to 0. We write: $\lim\limits_{x \to 0} g(x) = 3$.

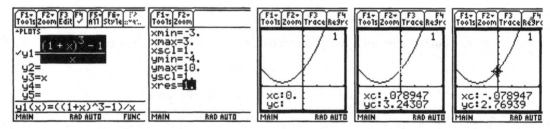

Figure 7.2 The trace cursor show that the function is undefined at xc:∅ *since the y-value is blank. The y-values on either side of the y-axis are close to 3; this is the graphical meaning of the limit.*

The numerical approach to limits

Now let's use the table approach in Figure 7.3 to see the same thing. Press TABLE and modify settings with TblSet (or F2 (Setup) from the TABLE screen) to see values close to zero. We get closer and closer by successively using ΔTbl=.1, .01, .001. At zero the function is undefined, as shown in the table.

x	y1
-.2	2.44
-.1	2.71
0.	undef
.1	3.31
.2	3.64
x=.2	

x	y1
-.02	2.94
-.01	2.97
0.	undef
.01	3.03
.02	3.06
x=.02	

x	y1
-.002	2.994
-.001	2.997
0.	undef
.001	3.003
.002	3.006
x=.002	

Figure 7.3 Table values closer and closer to zero, with y1 *values approaching 3.*

Tip: The graphic and numeric evidence may be very strong to indicate what the limit value should be, but don't make a speedy decision. Only by using careful mathematical analysis can you really prove that the limit exists.

An unreliable table

Remember that by setting INDEPENDENT to ASK in the Tb1Set window, you can specify table values. The table in Figure 7.4 is more succinct than the previous ones. The new function is $j(x) = \sin(2\pi/x)$. The first four lines of the table show even more persuasively that f approaches 1 and g approaches 3 as x approaches 0. The same reasoning suggests that

$$\lim_{x \to 0} j(x) = 0$$

The numeric data has led us astray. The function $j(x)$ is similar to the one used for the 'inaccurate graph' in Chapter 4; its graph hints that it has no limit. A value that is not the reciprocal of an integer, such as 0.003, shows that the function is not going to zero. The limit of $j(x)$ as x approaches 0 does not exist.

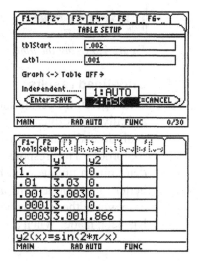

Figure 7.4 Selected values may be unreliable for guessing the limit.

The Limit function

The TI determines many, but not all, limits. The F3 (Calc) pull-down menu includes 3:limit. Although the command is entered linearly it is displayed in traditional mathematical format in the history area. In Figure 7.5 we find the two limits of the previous example. The syntax of the limit function is: is limit(*expression, variable, point*).

Local and long-run behavior

The limit can tell us about local behavior of functions. For example, one classic difference equation can be simplified as

$$f(h) = \frac{\sin(0 + h) - \sin(0)}{h} = \frac{\sin(h)}{h}.$$

Figure 7.5 The limit function.

We are interested in the local behavior close to zero; Figure 7.6 shows that the limit is 1. In the long run ($x \to \infty$) the limit is 0.

Also in Figure 7.6 we look at two rational functions, one where the limit is undefined and one where the limit is positive infinity at $x = 2$. Both functions have a vertical asymptote at $x = 2$. In the long run the limits are finite. One has a horizontal asymptote at $y = 1$ the other has a horizontal asymptote at $y = 0$.

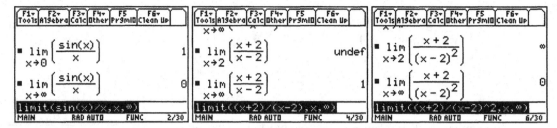

Figure 7.6 Using the limit to find local and long-run behavior for a function.

Undefined results

When given the result undef, take care in interpreting it as actually undefined. The TI uses several techniques, analytic and numeric, to do determine limits, but it is not foolproof. The response undef means that it did not determine a unique root, but one could exist, therefore it is better to interpreted this as "unknown." Figure 7.7 shows a case where the limit is not given until a variable constraint has been made. When *a* is constrained the limit is found.

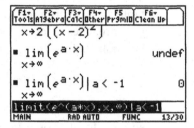

Figure 7.7 Limit with condition.

One-sided limits (optional)

There are some functions for which the limit is undefined at an *x*-value, but there is a finite limit if we approach from only one side. The classic example is a step function, where the value jumps at certain points. Such a function is *y* = int(*x*). The graph (Dot style) and limits are shown in Figure 7.8. Specifying a direction by adding a fourth entry to the limit function, it finds the requested one-sided limit (notice the superscript + or – in the pretty print format). Because the limits from the left and right are not equal, the limit as *x* approaches 1 (from both sides) does not exist. We also show that one of the rational functions from Figure 7.8 has limit -∞ as *x* approaches 2 from the left and limit +∞ as *x* approaches 2 from the right. Remember that one-sided limits might also be undefined; for example the limit of sin(2π/*x*) is undefined as *x* approaches zero from the right or left.

Figure 7.8 One-sided limits. The superscript + is hard to read because of the screen scale.

Tip: The direction value is read only as positive or negative — the magnitude is ignored. Use only 1 and -1 as the direction entries when computing a one-sided limit.

CHAPTER EIGHT

FINDING THE DERIVATIVE AT A POINT

The derivative at a point is the instantaneous rate of change mentioned in the previous chapter. In this chapter we learn several ways to use the TI to calculate the value of derivative at a point.

Using a graph to find the derivative at a point

If we zoom in for a microscopic view of a graph and it begins to appear linear; we are essentially seeing the line tangent to the graph. The slope of this tangent line is the derivative at the center point.

The derivative at a point on a graph is the slope of the tangent line

For a simple parabola, y1=x^2-3, in a ZoomDec window, we use the F5 (Math), A:Tangent to draw a tangent line to a graph at a point. You are prompted to select the point where you want a tangent line to be drawn: move the cursor to the point or enter its *x*-value and then press ENTER. See Figure 8.1.

The tangent line is drawn and its equation is given at the bottom of the screen. We see that the slope of the line tangent to the parabola at (1.1, -1.79) is 2.2, so this is the value of the derivative of the function at *x* = 1.1.

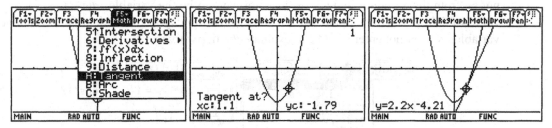

Figure 8.1 The tangent line to the graph y=x^2-3 at x=1.1. The tangent line slope of 2.2 is the value of the derivative at x=1.1.

Tip: If you want to draw a second tangent line, the first one remains on the screen unless you use F4 (Regraph) to start with a fresh graph.

The derivative at a point on a graph without the tangent line

Tangent lines are like training wheels on a bicycle: they eventually become unnecessary. We now directly find the derivative at a point. In Leibniz notation, the symbol for the derivative is *dy/dx*. This is the first selection of the F5 (Math), 6:Derivatives submenu available from the graph window. (Choices 2, 3 and 4 in the derivative submenu are available only when graphing in the 3D mode.) Figure 8.2 shows finding the derivative of the function that is the radius of a sphere in terms of its volume. Since the function may be new to you a graphic approach is a good idea.

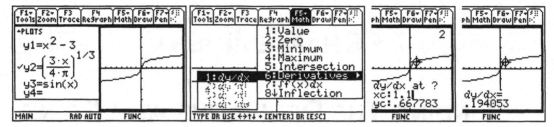

Figure 8.2 Using F5 (Math) 6:Derivative 1:dy/dx *to find the derivative at a point on the graph.*

Tip: Press the right arrow to activate a submenu, even though it may appear on the left.

Finding the derivative at a point without graphing

If the function is well known to you then it is safer to find the derivative at a point without a graph. From the HOME screen we can find the derivative value of y3=sin(x) at $x = 1.1$, by using the F3 (Calc), 1:*d* differentiate and completing the command as

$$d(y3(x),x)|x=1.1$$

The command doesn't look like standard calculus notation when typed in, but it translates to the traditional form in the history area. This is similar to using 3:limit which we saw in the previous chapter. The bar | is read as "such that" or "with." This adds the condition to specify the point at which we want the derivative value.

In Figure 8.3 two responses to the same entry are shown, but by pressing ♦_≈ instead of ENTER the second evaluation gives a decimal approximation of the answer. The function need not be from the Y= editor and also x does not have to be the variable: $d(t^2,t)|t=1.1$ gives a value of 2.2, just as we found for y1 above. Warning: you must identify the independent variable; you cannot enter y1'(1) to evaluate the derivative.

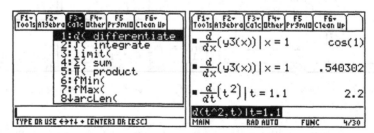

Figure 8.3 Using d to find the derivative of a function at a point.

Tip: The variable assignment after the "such that" bar applies only for this one command, in contrast to the more lasting effect of the entry 1→t.

Tip: To enter *d*(you may also use 2nd_*d* above the 8 key or use the CATALOG. In all cases beware that the italicized *d* is not the same as the lower case d.

Comparing the exact and the numeric derivative

The F3 (Calc) menu offers another derivative command, nDeriv, standing for "numerical derivative." Why two options? There are advantages and disadvantages to each.

Limitations of *d*

The symbolic derivative command is limited because it finds values using a formula-based approach, it knows that for $y = \sin x$, $y' = \cos x$. Be prepared for an error message if your function is out of the ordinary. For example, the function int is not on the known list, so the expression $d(\text{int}(x),x)|x=.5$ comes back unevaluated (it does replace int with the technical name floor). In such cases, we can use the numeric derivative command.

This numeric derivative is not limited by the calculator's compendium of symbolic derivatives and finds values for any function. For instance, we see in Figure 8.4 that nDeriv gives the correct value for the derivative of $y = \text{int}(x)$ at $x = .5$.

False results from nDeriv

The fact that nDeriv is an approximation can get us into deep trouble with certain points of some functions. For example, we know the function $f(x) = 1/x$ is not defined at zero and thus has no derivative there. We see in Figure 8.5 that using *d* to find the derivative at $x = 0$ gives -∞, i.e., the derivative does not exist. But nDeriv erroneously gives a value of 1 million. Neither of the graphic derivative methods described earlier give an answer from the graph of $y = 1/x$ when we enter $x = 0$.

The problem with nDeriv is not restricted to the obvious cases where the function itself is undefined. Our function $y3=(3x/(4\pi))^{\wedge}(1/3)$ is defined for all real values, so you might expect to find a derivative at $x = 0$ in the same way we did for $x = 1$. However, the tangent line at $x = 0$ is vertical. This means that the derivative is undefined there since a vertical line has no slope. Figure 8.6 shows this situation where, again, nDeriv gives a wrong answer and *d* responds with undef. The two graphical derivatives give No solution error messages when we enter $x = 0$ from the graph of y3.

Choosing between *d* and nDeriv

At worst, *d* is limited: it never gives an incorrect answer, but may respond with undef even though the derivative can be computed. Use nDeriv with caution when your function is unusual.

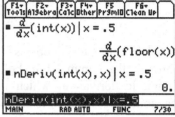

Figure 8.4. Using nDeriv when the symbolic integral is not known.

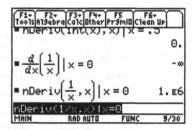

Figure 8.5 The home screen derivatives can give false results at an undefined point. But from a graph the derivative or tangent is not given at an undefined point.

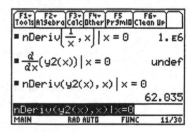

Figure 8.6 Invalid responses from nDeriv while d gives the correct answer.

CHAPTER NINE

THE DERIVATIVE AS A FUNCTION

Knowing the value of the derivative at a point, we define a new function called the derivative of f:

$$f'(x) = \lim_{h \to 0} \frac{f(x+h) - f(x)}{h}, \text{ if the limit exists.}$$

The computer algebra system of the TI gives us the derivative function in symbolic notation through the command d. For the comparably few functions beyond its recognition, use nDeriv (numeric derivative) which is correct "most of the time." But even with this strong symbolic capability, it is crucial to explore the graphs of a function and its derivative as a means to understand the relationship between a function and its derivative.

Comparing the numeric to the symbolic derivative

When the F3 (Calc) menu commands

$$d(\sin(x),x) \text{ and } nDeriv(\sin(x),x)$$

are evaluate without restriction they are functions themselves, see Figure 9.1.

Recall that d works from a list of known derivatives and differentiation rules while nDeriv uses a numeric method that is transparent from its output:

$$nDeriv(\sin(x),x) = \frac{\sin(x+h) - \sin(x-h)}{2h}$$

This is a version of the limit definition of the derivative with the default value $h = .001$ that produces the output in Figure 9.1. The h value is altered by adding a new value after the independent variable in the nDeriv command, so:

nDeriv(sin(x),x,.00001) evaluates as 50000.(sin(x+.00001)-sin(x)).

From this example we now understand the false result of the previous chapter:

$$nDeriv(1/x,x)|x=0 \text{ evaluates as } \frac{1/(0+.001) - 1/(0-.001)}{2(.001)} = 1.E6$$

As discussed in the previous chapter, use d whenever possible.

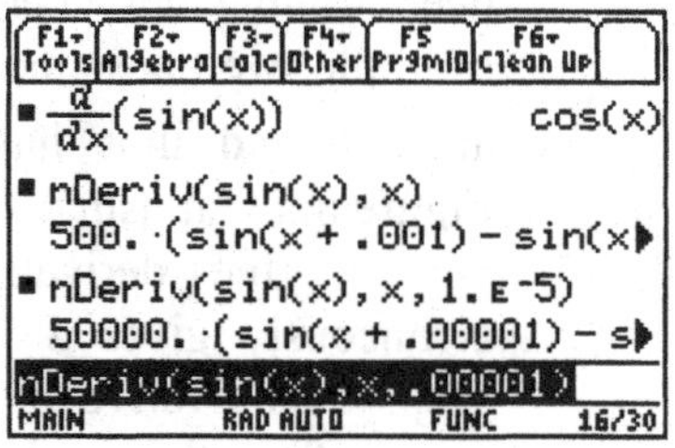

Figure 9.1 Contrasting d and nDeriv for symbolic and numeric derivatives.

Viewing a graph of a derivative function

Next we look at the graphical meaning of the derivative.

Matching a function to the graph of its derivative

We consider again the three classic functions from the last chapter:

$$f(x) = x^2 - 3, \quad r = h(V) = (3V/4\pi)^{(1/3)} \quad \text{and} \quad g(x) = \sin(x)$$

The three functions are stored in y1, y2, and y3 and graphed using a ZoomDec window. The graph of the derivative function for each of the three functions is displayed in Figure 9.2. Can you match the function to the graph of its derivative function?

We could, of course, graph them one by one, or use trace to identify them, but using a little thought we can identify them just by looking at the features of the graph. Consider the parabolic function f: it is decreasing until it reaches zero, then it is increasing. Since the derivative gives the instantaneous rate of change, this means that the derivative values are all negative to the left of the origin and are all positive to the right of the origin. Of the three options, this describes the line. (You may also know the power rule of derivatives that says the derivative of a quadratic function is a linear function.) Now consider the sine function: it oscillates between decreasing and increasing, so the derivative should oscillate between negative and positive. There is only one function that does this and it looks like a cosine function (another rule you may know). The remaining derivative function is always positive and has a spike at zero; this fits the slope patterns of the y3 graph. As a reminder, the derivative function of *h(x)* is undefined at zero; its graph has a "hole" there.

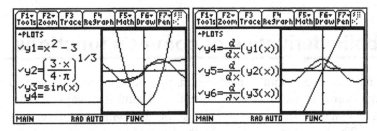

Figure 9.2 Matching derivative functions for three classic functions.

Graphing a function and its derivative in the same window

It is instructive to graph a function and its derivative in the same window, but you may want a means of distinguishing which is which. This is accomplished by using the F6 (Style) menu in the Y= editor. By choosing 4:Thick, you can change the graphing mode to a bold line. Unfortunately, there is no indicator of a function's style, save pulling down the Style menu for each function and looking for the option with the check to the left of its name. See Figure 9.3.

Functions and their derivatives do not usually fit very well in the same window. The slight modification of adding a constant to any function shifts its graph, while leaving the graph of the derivative unchanged. There is a lesson here. Showing the graph and its function in the same window is a parlor trick and must be carefully designed to work. Further, it should be realized that a modeling function and its derivative use different units on the *y*-axis. For example, when modeling motion, the distance function might be in feet and the derivative function would then be in feet per second.

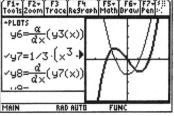

Figure 9.3 Two styles to distinguish a function and its derivative function in the same window.

The function that is its own derivative: $y = e^x$

Could there be some function that is not changed by taking the derivative? In other words, could some function be its own derivative? To save time in guessing, we try an exponential function. In Figure 9.4, we define y9 to be our old doubling function $y = 2^x$ in bold style and graph it along with its derivative y10.

Now change y9 to the tripling function $y = 3^x$ and graph again. The graph of this function is very close to the graph of its derivative. The exponential function we are looking for has a base between 2 and 3. In calculus, we find that this amazing number is 2.718..., an irrational number denoted by e. Make a table or use Trace to confirm that the two functions are equal.

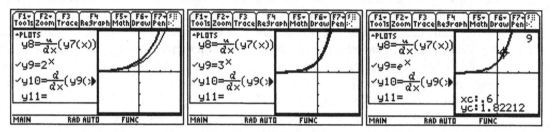

Figure 9.4 Looking for a function that is its own derivative.

The symbolic derivative of common functions

This brings us to the point where we trumpet the symbolic power to calculate derivatives. From the home screen you can request, and almost always get, the symbolic form of a derivative function. In Figure 9.5 we show the symbolic derivatives for several common functions. The history area looks like the list that appears in the back cover of many popular calculus books. Although it is reassuring to know they can be pulled up quickly, you should know these formulas by heart.

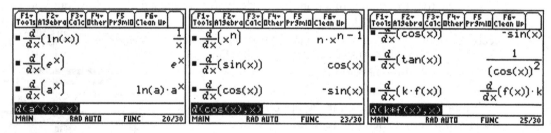

Figure 9.5 A gallery of famous derivatives.

CHAPTER TEN

THE SECOND DERIVATIVE: THE DERIVATIVE OF THE DERIVATIVE

In Chapter 9 we saw that the derivative itself is a function. Considering the derivative as a function, there is nothing stopping us from finding *the derivative of the derivative*. This is called the second derivative of the original function $f(x)$, written $f''(x)$. We find that the second derivative tells about the concavity of a graph.

The second derivative syntax

The TI differentiation command d has a third optional input that allows us to take second (and higher) derivatives. The full syntax is

$$d(function, variable\,[,order])$$

and the default setting for *order* is one if the order is not entered. We enter $d(f(x),x,2)$ for $f''(x)$. Figure 10.1 shows the second derivatives of two common functions and the pretty print version of the input. A superscript of 2 is displayed in both the numerator and denominator of the second derivative notation.

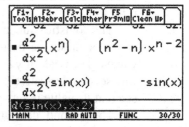

Figure 10.1 Second derivatives of common functions.

Symbolic derivatives and their graphs

Consider the simple function $y = x^2-3$. We enter it as y1 and define y2 and y3 to be its first and second derivatives. In Figure 10.2, the GRAPH screen (using ZoomDec) shows the three graphs, using Thick for y1. The line through the origin is the graph of y2, giving the rate of change of the parabola y1. Considering y3 as the derivative of y2, the rate of change is simply the slope of the y2 line, so y3 is the horizontal line $y = 2$. The trace mode identifies the second derivative.

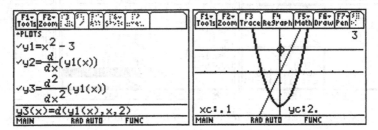

Figure 10.2 The graph of $y = x^2-3$ and its first and second derivatives.

The second derivative of y1, tells us the concavity of the graph of y1; the second derivative is positive (a constant of 2), so the graph of y1 is concave up.

> *Tip:*   If the graph is taking a very long time, press ON to stop it and reset xres to improve the speed.

Looking at the concavity of the logistic curve

In general, trying to graph a function and its derivative in the same window is not practical. This is even more of a challenge with the first and second derivatives. We use three different window settings in Figure 10.3 to analyze a logistic function and its derivatives. The domain for x is the same in each screen, but we change the range values to best display the behavior of each function. The logistic function was entered as

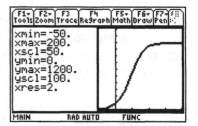

$$y1= 1000(1+9e^{\wedge}(-.05x))$$

There is no need to change y2 and y3; since they already refer to y1.

The logistics function is monotonically increasing on the interval so the derivative function y2 has all positive values. The peak of y2 is the point of fastest growth of y1. The second derivative y3 is the rate of change of y2 and we see that its value is zero at the peak of y2.

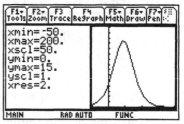

The second derivative, y3, is zero at about $x = 40$; which, on the graph of y1, is the point where the concavity is changing from up to down. We call this a point of inflection on the graph. When the second derivative is positive the graph of the function is concave up, when it is negative the graph is concave down. Since the point of inflection is an important point on y1, we want to identify it more exactly.

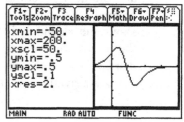

Figure 10.3 The logistic function and its two derivative functions.

The inflection point from the graph

Our way of identifying the point of inflection is to find the zero of the second derivative. With the second derivative graph in the window, use F5 (Math), 2:Zero and find the zero as we did in Chapter 5. The final computation is shown in the first frame of Figure 10.4.

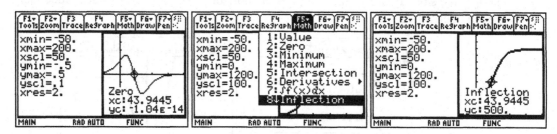

Figure 10.4 Using 2:Zero to find the zero of the second derivative (remember that the y-value may be only extremely close to zero).

A more direct way to find an inflection point of the logistic function is to work from the graph of the original function, y1, and use F5 (Math), 8:Inflection. The last two frames in Figure 10.4 show this approach and that either way we obtain the same value for the inflection point, $x = 43.9445$.

When you graph y3 you may notice that it graphs much more slowly than you are accustomed to. Even moving left and right with the trace cursor requires some patience. This is because y3 is not drawn from its own formula, rather from an operation on y1. Were we to put in the formula for y3, it would graph much more quickly.

Creating a numeric second derivative using a table

We reexamine the data from the beginning of Chapter 7 giving the height of a grapefruit tossed into the air over six seconds. Recall that we stored variables and functions in long names so that, unless you erased them using 2nd_VAR-LINK, the information is still there. If not, you can simply reenter it now.

In Figure 10.5 we first recalculate the average velocity computation giving a list starting with 84. The derivative of velocity (the second derivative of height) is acceleration. We compute the average acceleration from the original data by computing ΔList of the average velocities divided by 1, the time intervals. Notice that we do not need to give the average velocity list a name; it has the temporary name ans(1) when it is the most recent calculation in the history area.

The list produced at the end of Figure 10.5 has five identical entries of -32. This number may be familiar to you as the downward acceleration due to gravity (measured in feet per second squared). Computing the second derivative from this tabular data of heights reminds us of the constant effect of gravity.

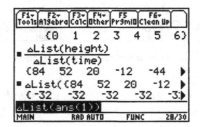

Figure 10.5 Computing average accelerations from a list of velocities.

CHAPTER ELEVEN

THE RULES OF DIFFERENTIATION

Using the definition of the derivative to find a derivative function is cumbersome; fortunately there are shortcuts to finding derivative functions. We use the calculator to show the main rules and see examples. These rules must be proved analytically, but a graphical study steers us away from common errors.

The Sum Rule and the Constant Factor Rule

The constant factor rule, $f'(k \cdot x) = k \cdot f'(x)$, is shown in dy/dx notation in Figure 11.1. The TI symbolic manipulator factors the k to the right side. Next the sum rule for functions $(f(x) + g(x))' = f'(x) + g'(x)$ is shown in symbolic terms. The second frame of Figure 11.1 shows how the TI applies these rules when given actual functions. When applying the sum rule the order of output may be reversed as shown.

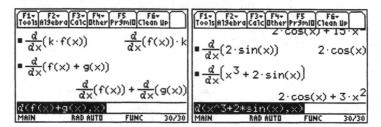

Figure 11.1 Two differentiation rules and examples.

Tip: A symbolic answer may be written in any number of equivalent ways.

The Product Rule

The product rule, $(f(x) \cdot g(x))' = f(x) \cdot g'(x) + g(x) \cdot f'(x)$ is seen in Figure 11.2. It requires scrolling to see the final f(x) at the end of the symbolic output.

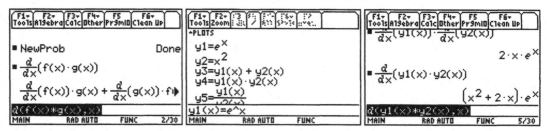

Figure 11.2 The Product Rule and an example to show (f·g)' ≠ f'·g'.

Let's suppose we make a tempting guess that the derivative of the product is the product of the derivatives. This is not unreasonable since the derivative of a sum is the sum of the derivatives, but it is wrong. In the second frame of Figure 11.2 we use y1=e^x and y2=x^2 to test the product rule. First we find the product of the derivatives and then the derivative of the product function. They are definitely not the same.

The Quotient Rule

Let's check the quotient rule in the same way. First we generate the formula in Figure 11.3, but, as is common with computer algebra systems, the answer is not in the traditional form. The F2 (Algebra) command 6:comDenom gives the symbolic derivative as the familiar single fraction. In the final frame we find the derivative of the quotient y1=e^x and y2=x^2. Since the quotient of the individual derivatives is easily calculated as (e^x)/(2x) we see that in general, $(f/g)' \neq f'/g'$.

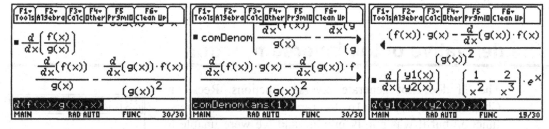

Figure 11.3 The Quotient Rule put in a familiar form by rewriting the symbolic derivative as a single ratio using the comDenom command. And, an example to show $(f/g)' \neq f'/g'$.

The derivative of the tangent function as a quotient

The tangent function has a quotient formula definition $\tan(x) = \sin(x)/\cos(x)$; the quotient rule gives $y' = 1/\cos^2(x)$. In Figure 11.4 we verify this derivative. The derivative expression was simplified in the numerator by using the Pythagorean Identity. We show this identity by entering it on the entry line. The output `true`, means that the equation is true for all x.

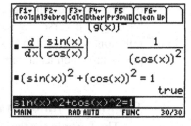

Figure 11.4 The derivative of the tangent function.

Tip: The conventional mathematical way of writing a power of a trigonometric function, such as cos²(x), cannot be used on the TI. Instead, use cos(x)² or, better yet, (cos(x))².

The Chain Rule

Thinking of a function as a composite function and then using the chain rule often simplifies finding the derivative. Consider $y = (x^2+1)^{100}$. A straight-forward but impractical approach is to expand the expression, (write it as a polynomial of degree 200), and then differentiate term by term. Instead, we apply the chain rule and find the derivative quickly and easily:

$$y' = 100(x^2+1)^{99}(2x) = 200x(x^2+1)^{99}$$

Figure 11.5 shows an unsuccessful attempt to show the general chain rule formula using undefined functions f and g. However, once we define f and g, the chain rule is properly applied. We also verify the answer by finding the derivative of the composed function by entering it directly.

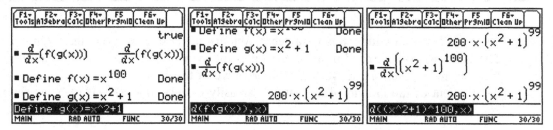

Figure 11.5 The Chain Rule cannot be shown in abstract form, but works for specific functions.

The derivative of the inverse function

The TI does not generate inverse functions. Recall in Chapter 5, we used DrawInv to graphically show an inverse function but it was purely graphic and we were unable to trace or evaluate the inverse function from the graph. However, there are some inverse functions whose formulas are built-in; these include the trigonometric functions and the hyperbolic functions. Figure 11.6 shows the evaluation of two inverse functions. There are two conventions for writing an inverse trigonometric function; the TI uses the $^{-1}$ notation, such as $\sin^{-1}(x)$. The alternate way is to use the prefix arc, such as $\arcsin(x)$.

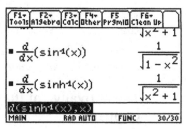

Figure 11.6 Derivative of inverse functions.

The inverse function formula

The formula for the inverse function formula,

$$\frac{d}{dx}(f^{-1}(x)) = \frac{1}{f'(f^{-1}(x))}$$

can be difficult to understand. But using a real example can help clarify its meaning. The function $f(x) = x^3$ and its inverse $f^{-1}(x) = x^{1/3}$ have derivatives as shown in Figure 11.7. The relationship is then seen as

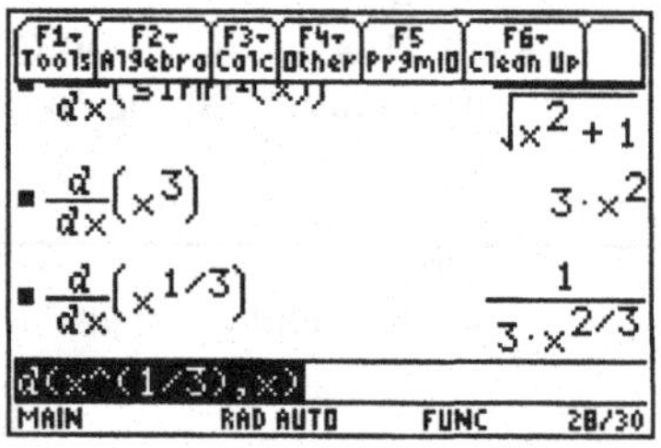

Figure 11.7 Derivative of the inverse relationship.

$$\frac{d}{dx}(f^{-1}(x)) = \frac{1}{f'(f^{-1}(x))} \text{ becomes } \frac{d}{dx}(x^{1/3}) = \frac{1}{3(x^{1/3})^2}$$

Implicit differentiation (Optional)

There are no built-in implicit differentiation capabilities. However there are APPS that enhance the TI. The screens shown are from the APP *Calculus Tools*. This APP is a free download from at the *education.ti.com* site. The CD provided with your calculator guides your downloading of this software when you are connected to the internet.

After being downloaded, Calculus Tools appears on the APPS Desktop. Highlight the icon and press ENTER. If you press APPS and a menu appears instead of the desktop, choose FlashApps and Calculus Tools from the submenu. The Apps Desktop is turned ON or OFF in the MODE menu, Page 3.

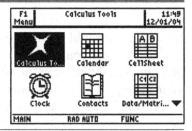

The home screen shows a set of tabs that are pull down menus. The applications are selected from F2 to F6. The F1 (Tools), ANSWER option is often used after an applications output, since the output does not always fit on a single screen.

Use F2 (Deriv), 4:Implicit Difn to start the application.

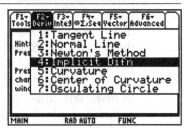

A sample example appears as the default for each application. In this case, we are finding the derivative of the equation describing a circle. Initially the entry mode is set to ALPHA, so in order to change to a second order derivative you must press ALPHA (to leave alpha mode) before entering 2.

Press ENTER and the implicit differentiation is shown. In this example the answer is simple enough to fit on the screen and there is no need to use the F1 ANSWER option.

CHAPTER TWELVE

OPTIMIZATION

One of the powerful uses of the derivative function is to find the maximum or minimum of a function and to find the concavity of a graph. But we must confess that calculators and computers with graphing capabilities can, in most cases, find maximum and minimum values of a function without your having to know anything about calculus. In this section there are examples showing both the calculus and the non-calculus approaches.

The ladder problem

Typically, optimization problems arise from real-world applications. The ladder problem is to determine the longest ladder that can be carried horizontally around a corner that joins two hallways. We assume that the hallways are different widths: the narrower one is 4 feet wide, the wider one 8 feet. Figure 12.1 shows the position where the ladder could get stuck: it touches both walls and the corner. For the ladder to make this corner, it needs to fit in the hallway for every angle θ, even the tightest one. Think of the ladder as being divided at the corner point and use the triangle trigonometry definitions for the sine and cosine to find each piece. This leads to

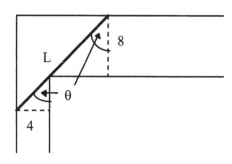

Figure 12.1 A ladder carried horizontally around a corner.

the following equation for the ladder's length in terms of the angle. We want to <u>minimize</u>

$$L = f(\theta) = \frac{4}{\sin(\theta)} + \frac{8}{\cos(\theta)}$$

A ladder of that minimal length fits for all angles, and it is the longest possible ladder that works since it touches both walls at the tightest angle.

Finding an equation is the hard part of an optimization problem. We now show several different methods for finding the angle value that minimizes L. Before beginning, check the status bar for RAD to be certain that the angle mode is set to radians.

Graphic solutions without calculus

Enter L as y1=4/sin(x)+8/cos(x). You might be tempted to use a ZoomTrig setting, but, like most models, there is a more restricted domain in this example. The angle must be greater than 0 but less than $\pi/2$. To set the window y-values, we are generous and say that the minimum ladder is under 60 feet (we also set ymin=-10 to avoid any overlap with the numeric values at the bottom of the screen).

We use F5 (Math), 3:Minimum to directly find $y \approx 16.6478$. As shown in Figure 12.2, you provide a lower bound and an upper bound and it finds the minimum. The bound values are entered directly; recall the *y* must be within the viewing window, and that the arrow keys can also be used to set bounds.

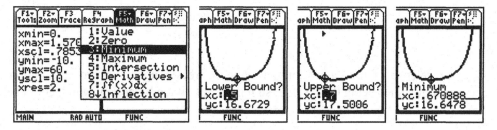

Figure 12.2 A minimum found on a graph by using F5 (Math) 3:Minimum.

Calculus solutions

The same results can be found symbolically with calculus. From the HOME screen, Figure 12.3 shows the calculus approach: find the zeroes of the first derivative in the restricted domain, evaluate the critical value in the original function, and apply the second derivative test to verify that the extremum is a minimum. Notice how we can use ♦_θ as a variable (the θ key is above the ^ key). In general, this use is not recommended since it takes two keystrokes each time it is entered. The first entry gives multiple solutions so we must restrict the domain to $0 < \theta < \pi/2$, as shown in the second entry. In the final frame the screen shows the evaluations were entered using ans(1) and ans(2) respectively.

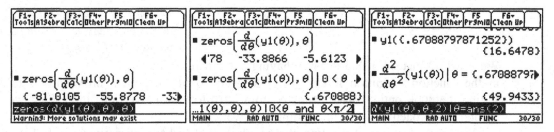

Figure 12.3 A calculus solution for minimum ladder length.

Why bother with calculus?

We always hope for an exact answer, which only derivatives can lead us to. Such "closed form" solutions can be important for using in other calculations or for insuring unlimited accuracy.

The TI did not give us an exact answer for this problem because it was unable to symbolically solve for the zero of the derivative. This is common, particularly when trigonometric functions are involved. Figure 12.4 shows the TI derivative as a fraction. This is zero only when the numerator is zero. We can solve for *x* by hand using algebra and find the exact answer.

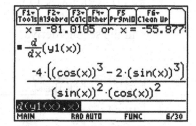

Figure 12.4 The derivative as a fraction.

$$-4(\cos(x)^3 - 2\sin(x)^3) = 0$$

$$-4(1 - \frac{2\sin(x)^3}{\cos(x)^3}) = 0$$

$$\tan(x)^3 = \frac{1}{2}$$

$$x = \tan^{-1}(\sqrt[3]{\tfrac{1}{2}})$$

In defense of the calculator, its symbolic reasoning makes no assumptions about the domain of values being $0 < x < \pi/2$, so it does not divide by $\cos(x)$, which could be zero.

Box with lid

Suppose we have an 8.5 by 11 inch sheet of paper and want to cut squares and rectangles from the corners to create a folded box with lid. See Figure 12.5. We want to maximize the volume. Notice that if the x cut is very small, then the box is so shallow that it hardly holds anything. If the x cut is large, then the bottom is so small that the box again holds very little. We first find a volume function that depends only on the length x of the cut-out square:

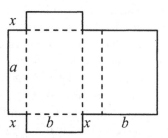

Figure 12.5 A diagram of cutting corners to create a box with lid. Fold on the dotted lines.

$$V = a \cdot b \cdot x = \left(\frac{17}{2} - 2x\right) \cdot \left(\frac{11 - 2x}{2}\right) \cdot x$$

We use 17/2 instead of 8.5 because having a decimal in the function definition forces all TI answers to be decimal approximations (although setting the MODE option Exact/Approx from AUTO to EXACT overrides this). First we take a graphical approach to see an approximate answer.

We graph the volume function along with its first and second derivatives. In Figure 12.6, we enter the function $V(x)$ as y1. To avoid slow graphing, compute the derivatives on the HOME screen and use copy-and-paste (♦_C & ♦_V) to put them into the Y= editor. The functions have been graphed using three different styles in the window shown. From the graph, press F3 (Trace), move to the second function (down arrow), and find an approximate zero of the derivative. Notice that at the given x-value the volume function appears to be at a maximum and the second derivative is negative. This graphically confirms the calculus theory.

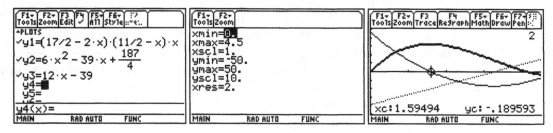

Figure 12.6 Using a graph to find the maximum of a function by tracing the first derivative.

To find the maximum analytically from the HOME screen, we use fMax in Figure 12.7. This example shows that we need to know something about the function and the model's domain, since an unrestricted entry of fMax gives x=∞. The function is cubic so it has both a local maximum and a local minimum. When restricted to 0 < x < 4, a single analytic answer is shown. The exact answer is then shown in decimal approximation form using ♦_≈. Finally, the second derivative test confirms that the point is a maximum as desired.

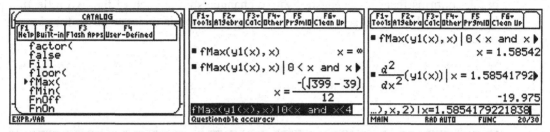

Figure 12.7 Using fMax to find the maximum of a function and checking with the second derivative.

Tip: Ideally, you look at the graph of the function and its derivatives before proceeding to the symbolic computations.

Tip: For entering conditions, use < and > whenever possible since they are available directly on the keypad.

Normal density family: using f" to find concavity

An important family of functions in statistics is the *normal density* function. It has a graph that is bell-shaped and a definition related to the family of functions

$$f(x) = e^{-(x-a)^2/b}$$

To study this family, we enter the definition and later set the parameters a and b as needed.

We discussed in the last chapter that the second derivative measures the concavity of the graph. Points where concavity changes from up to down, or vice versa, are of particular interest. Where the second derivative changes sign it has a zero. In Figure 12.8, we enter the general form into y1 and clear the variables and find the zeros of the second derivative. The two solutions depend on *a* and *b*. Setting $a = 0$ and $b = 1$ results in inflection points at x= ±1. By graphing other values of a, you can see that this parameter controls the horizontal shift of the graph. Setting a = 0 makes the graph centered at the origin.

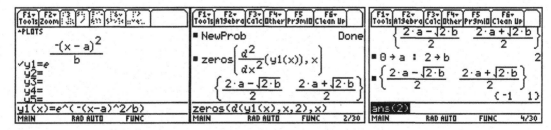

Figure 12.8 Finding the two inflection points of a bell-shaped curve.

Just as all critical points of a function are not necessarily relative extremum, there can be zeros of the second derivative function which are not points of inflection. To check a point of inflection, it is normally easiest to consult the graph of the second derivative or check the concavity on either side of each zero (like the first derivative test but using the second derivative). In Figure 12.9, the function and its second derivative have been graph on the same screen and F5 (Math) 8:Inflection was used to find the inflection point at $x = 1$. This graph confirms that the second derivative changes sign at each of the two inflection points. Further, it makes visual sense that the original function is concave down for $-1 < x < 1$ and concave up otherwise.

Tip: There is no built-in inflection point command that can be used from the HOME screen. Calculus techniques are required unless you use a graph.

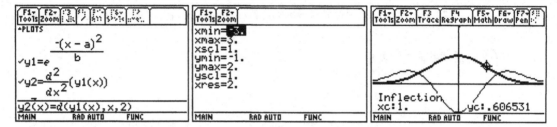

Figure 12.9 Finding the inflection point of a bell-shaped curve.

Part III Integral Calculus

Chapter Thirteen

Left-And Right-Hand Sums

The fundamental activity of integral calculus is adding. In the discrete case, we sum a set of values. In the continuous case, we use the integral to sum over an interval. In this chapter we restrict our attention to finite discrete sums. These sums are approximations to the value of a definite integral; we make that connection in Chapter 16, Riemann Sums.

Distance as the sum of the velocity data

Time (sec)	0	2	4	6	8	10
Velocity (ft/sec)	20	30	38	44	48	50

Table 13.1 Velocity of a car, every two seconds.

If you drive 50 miles per hour for 3 hours, then you have traveled 50 + 50 + 50 = 150 miles. We rarely travel a constant speed. If you drive 20 mph for two hours, 30 mph for the next two hours, and finally 40 mph for two more hours, then you have traveled 20(2)+ 30(2) + 40(2) = 180 miles over the six hours.

Table 13.1 shows velocity readings at 6 different times. We do not know if we traveled mostly at 20 ft/sec or 30 ft/sec for the first two seconds. If we assume the velocity is constantly increasing, then these two numbers give us lower and upper bounds for speed in the first two seconds. To get a lower bound on the distance traveled in the ten seconds, we reason as follows. For the first two seconds, the velocity was at least 20 ft/sec, so the car traveled at least 40 feet in that interval. In the second two seconds, the car traveled at least 60 feet. In general, a lower bound on the total distance traveled is a sum of the first five velocities multiplied by two. For an upper bound, double the sum of the last five velocities.

Creating and summing lists from the home screen

We enter the lower bound data in a list called `lo`. For the higher bound data list, `hi`, edit the first list when it is still on the entry line. See Figure 13.1. The `sum` command can be typed, or pasted from the `CATALOG`. We see the car traveled between 360 and 420 feet.

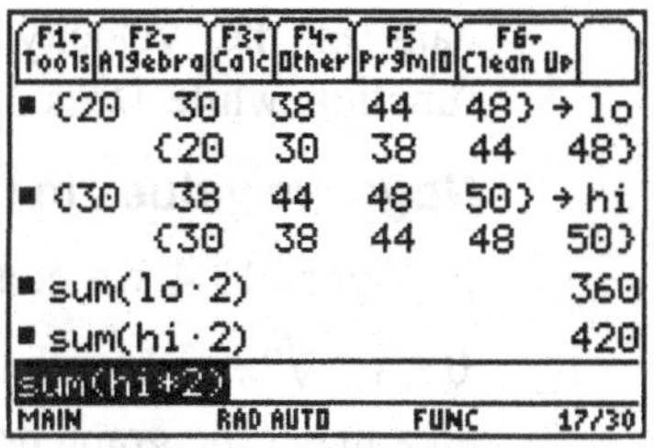

Figure 13.1 Finding lower and upper bounds on distance traveled.

Summing lists to create left- and right-hand sums

The previous example summed a simple data list. A common calculus task is to form the left- and right-hand sums for function values over an interval that has been divided into n equal subintervals (where n is an arbitrary integer that tends to get larger and larger). This is only slightly more complicated than what we have done: it is just a sum of products of function values and the fixed length of an interval subdivision. Geometrically, it is the sum of areas of a collection of rectangles that are $f(x)$ high and Δx wide. That is, it approximates the area between the function's graph and the x-axis. In Chapter 16 we use a program that draws the rectangles and calculates left- and right-hand sums; here we just want to understand how to build these sums.

The left- and right-hand sums

We first create the left-hand function sum and then make minor adjustments for the right-hand definition. We assume that the function whose values we sum is defined as y1. Sums are calculated over the interval $a \le x \le b$ divided into n of partitions, each of length $\Delta x = (b-a)/n$. The left-hand sum uses the function values at the left of each sub-interval, so the relevant x-values do not include b and are

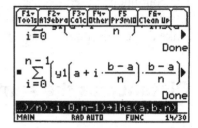

Figure 13.2 The left-hand sum defined as 1hs(a,b,n). Scroll to see all.

$$a, a + (b-a)/n, a + 2(b-a)/n, \ldots, a + (n-1)(b-a)/n.$$

We name the left-hand sum function 1hs and use the three letters a,b,n as independent variables. The entry of the definitions is too long to show on the calculator screen in Figure 13.2; they are:

$$\Sigma(\text{y1}(a+i*((b-a)/n)),i,\emptyset,n-1))*((b-a)/n),i,\emptyset,n-1)\rightarrow\text{1hs}(a,b,n)$$

$$\Sigma(\text{y1}(a+i*((b-a)/n)),i,1,n))*((b-a)/n),i,\emptyset,n-1)\rightarrow\text{rhs}(a,b,n)$$

The right-hand sum differs by using the function values at the right of each sub-interval, so we start at $a + (b-a)/n$ and go to $a + n(b-a)/n = b$, i.e., the index of rhs runs from 1 to n instead of 0 to $n-1$. Once entered, the formula can be tested by using the values shown in Figure 13.3. The MODE was changed from AUTO to APPROX answers. (Or you can press ♦_≈ each time.) Notice that the right-hand sum for $1/x$ is less than the left-hand sum. The data in the velocity example indicated an increasing function, while $1/x$ is decreasing.

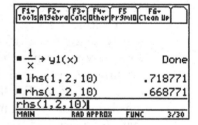

Figure 13.3 Testing 1hs and rhs.

Negative values in the sum

In Figure 13.4 we consider the function $f(x) = \sin(x^2)$. The left-hand sum on the interval $0 \le x \le \sqrt{2\pi}$ is less than the left-hand sum on the shorter interval $0 \le x \le \sqrt{\pi}$. How can the area under the graph decrease by using a larger interval? The graph shows that the function is negative when $\sqrt{\pi} \le x \le \sqrt{2\pi}$. The decrease in the sums now makes sense since we are adding negative values for rectangles in that interval Thus we need to be careful when interpreting these sums as areas. To find the actual area between the graph and the x-axis, we would have to use the absolute values of any negative terms in the sum.

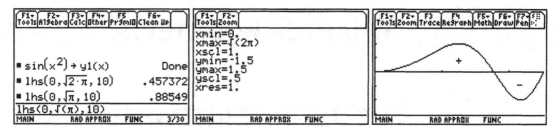

Figure 13.4 The total sum over an interval may be less than the sum over a subinterval if some of the function values are negative.

Tip: If you forget to press ♦_≈ and get an unexpected exact answer after a long calculation, you can use ans(1) ♦_≈. This saves the recalculation time. If you repeatedly need approximate answers, change the MODE to APPROX.

Approximating area using the left- and right-hand sums

By increasing the number of partitions, the left- and right-hand sums may approach a limit which we interpret as the (signed) area under the function's graph. We write this as

$$\int_a^b f(x)\,dx = \lim_{n\to\infty}\sum_{i=1}^n f(x_i)\cdot \Delta x$$

As an example, let's see if there is a limit to the left-hand sums of $y = \sin(x^2)$ over the interval $0 \le x \le \sqrt{\pi}$ as the number of partitions increases. The computations in Figure 13.5, with $n = 10$, 50 and 100, suggest that the sums approaches a limit. The TI has a built-in function nInt that reports the numeric limit in these cases.

We know from Chapter 7 that the TI has a Limit function, so we apply it to this left-hand sum. Unfortunately, we see in the second frame of Figure 13.5 that it does not evaluate the limit of the lhs function, although it does simplify the summing expression by setting $a = 0$ and $b = \pi$. There are infinite sums that the TI knows and these can be used to verify that the definite integral is the limit of the left- (or right-) hand sums. Using y1(x)=x^2, we see that both sums have the same limit on the interval $0 \le x \le 1$. In Chapter 15, we confirm that this number is the value of the definite integral by using the Fundamental Theorem of Calculus.

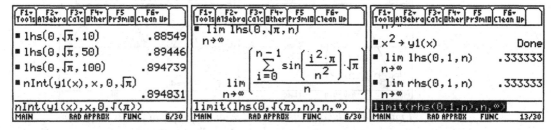

Figure 13.5 The left-hand sum for an increasing number of partitions tends to a limit. The limit command works in some but not all cases involving lhs and rhs.

CHAPTER FOURTEEN

THE DEFINITE INTEGRAL

In the previous chapter we calculated the left- and right-hand sums to approximate the signed area under a curve. The definite integral is defined as the limit of the left-hand (or right-hand) sum as the number of partitions n goes to infinity. Thus, each definite integral is a specific real number, and the TI calculates this value. (Well, almost — it calculates an approximation that is generally reliable.) The definite integral is evaluated as a number, but defining its upper limit as a variable creates a new function.

The definite integral from a graph

As with the derivative, values of the integral of a function can be found on both the HOME and the GRAPH screen. From a function's graph, we can find its definite integral *and* see the graphic representation. We start with $y = 2\sin(x)$ and graph it in Figure 14.1. Similar to finding the derivative from a graph, we use the F5 (Math) pull-down menu and its integral option 7: $\int f(x) dx$. We are prompted to set the lower limit and then the upper limit. These limits are set in the same way that you have already set bounds using several other commands. Remember that the limits must be within xmin and xmax.

The number $\int f(x) dx$ can be interpreted as the area of the shaded region. It may surprise you that the result is an integer, exactly 4.

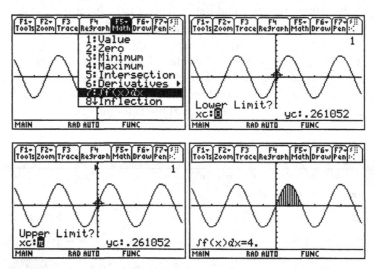

Figure 14.1 The definite integral from the graph of $y1 = 2\sin(x)$ over the interval $0 \le x \le \pi$ and using a ZoomTrig setting.

Tip: If there is more than one function graphed, then you must choose the desired function after selecting the $\int f(x) dx$ command, but before specifying a lower limit.

The definite integral as a number on the home screen

To find the same result without using a graph, you can work from the HOME screen. The TI command for the definite integral is the integral sign, the 2nd version of the 7 key. It is also a menu choice of for F3 (Calc). The entry line command in Figure 14.2 shows the syntax used to produce the answer along with the traditional mathematical symbolism in the history area.
The syntax is

$$\int(\texttt{function,variable,lower,upper}).$$

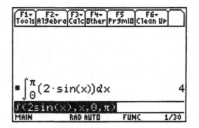

Figure 14.2 Integration from the HOME screen.

Facts about the definite integral

Four definite integral facts are illustrated below using simple functions and their graphs. You are encouraged to change the function and window to make them more exciting. The following examples show the result graphically and numerically. It is important that you feel comfortable using both methods of finding the definite integral.

> *Tip:*   After a graph with shading has been displayed, clear the screen with F4 (ReGraph) before the next graphing.

Reversed limit integrals are the negative of one another

Unlike most parameters, which prompt error messages whenever xmin > xmax or Lower Bound > Upper Bound, the Upper Limit and Lower Limit can be in either order. Reversing the order changes the sign of the answer, as demonstrated in Figure 14.3. A graph method is not shown here because shading does not indicate the reversal of the limits The $\int f(x)dx$ value does switch sign.

The intermediate stop-over privilege

The definite integral can be calculated as a whole from the lower to upper limit, or it can be calculated in contiguous pieces. This can be thought of as a plane fare where the charge is the same whether you fly non-stop or have intermediate landings. In Figure 14.4 we get the same answer as before by considering our function y1=2sin(x) over $0 \le x \le \pi/3$ and then $\pi/3 \le x \le \pi$.

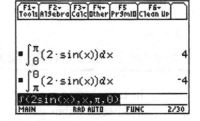

Figure 14.3 Limit reversal changes the sign of the definite integral.

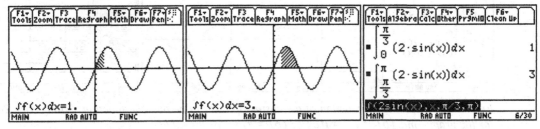

Figure 14.4 A definite integral found in two pieces. The total, 3+1=4, is the value across the whole interval.

> *Tip:* There are four styles of shading that are used consecutively, so your shading styles may differ from the ones shown in the examples.

The definite integral of a sum is the sum of the integrals

In Figure 14.5, we have set y1=2sin x and y2=x. The graph of the function y3=y1(x)+y2(x) is shown as a Thick curve. We see, both graphically and symbolically, that the definite integral of the sum function y3 is the sum of the definite integrals of the two summand functions.

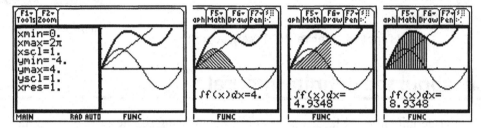

Figure 14.5 Example of the definite integral of a sum is the sum of the definite integrals.

Constant multiples can be factored out of a definite integral

We saw an example of this fact earlier with $\int_0^\pi 2\sin(x)dx = 2\int_0^\pi \sin(x)dx$. The symbolic rule for this and for the sum rule is shown in Figure 14.6. Notice that the constant is factored to the right side of the answer, while the traditional factoring is to the left. Other examples can be derived with actual functions and real number limits.

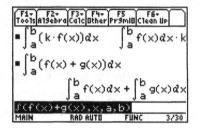

Figure 14.6 A constant multiple is factored out of a definite integral. And the sum rule.

The definite integral as a function

In this chapter's integral examples, we have used x as the function variable. We could have used some other dummy variable in the definite integral command and the numeric answer would be the same. For example, $\int(\sin(t),t,\emptyset,\pi)$ would also give 2.

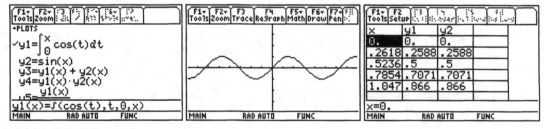

Figure 14.7 The integral function of the cosine appears to be the sine function. Settings are ZoomDec for the graph, and tblStart = Ø and Δtbl = π/12 for the table.

We usually think of the definite integral as a number resulting from the command ∫f(t),t,a,b). If we change b, we get a new value for the integral, so ∫f(t),t,a,b) is then a function of b! As an example, use cos(*t*) as the function, choose 0 for the lower limit, and in the Y= editor define

$$y1 = ∫(cos(t),t,0,x)$$

You may be wondering how we can do this: doesn't *x* have to be the variable for functions in the Y= editor? The *t* in the integral is a dummy variable; the answer to the definite integral depends on *x*, so y1 ultimately is a function of *x*. The graphing shown in Figure 14.7 is unbelievably slow because of the intensive numerical work to calculate each definite integral. The graph of y1 looks like a sine function. In the last frame of Figure 14.7 we compare table entries of y1 to y2=sin(x). You could also graph the sine function and check that the two functions have the same graph.

Tip: Graphing functions defined with ∫(..,x) is quite slow; setting xres to a higher value increases graphing speed.

The TI computer algebra system can also verify that the graph in Figure 14.7 is the sine function. On the home screen in Figure 14.8, the entry ∫(cos(t),t,0,x) gives sin(x). We also see that interesting things happen when we change the lower bound from zero to π/2 and then on the entry line 3π/2. The last entry produces sin(x)+1. The three functions defined this way differ by constants, so graphically they are vertical shifts of one another. Avoid graphing from definite integral definitions – life is too short.

If you think of the derivative as the rate of change of *y*-values, then these three should have the same derivative. Each of these three functions is called an antiderivative of cos(*x*). The common notation used is

$$\int \cos(x)dx = \sin(x) + C$$

where *C* is an arbitrary constant. This is the focus of the next chapter.

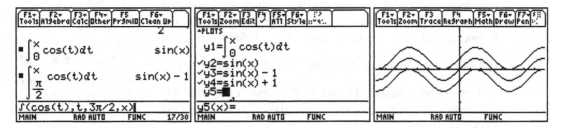

Figure 14.8 Integral functions of the cosine starting with different initial values are of the form sin(x) + C. Graphing functions with known formulas is much faster than using an integral definition.

CHAPTER FIFTEEN

THE FUNDAMENTAL THEOREM OF CALCULUS

The Fundamental Theorem of Calculus is discussed in two forms: as the total change of the antiderivative, then as a connection between integration and differentiation.

The Fundamental Theorem

The Fundamental Theorem of Calculus states:

if f is a continuous function and $f(t) = \dfrac{dF(t)}{dt}$, then

$$\int_a^b f(t)\,dt = F(b) - F(a)$$

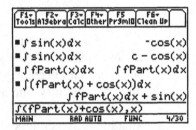

Figure 15.1 Two forms of finding the antiderivative. When the antiderivative cannot be found, the integral notation shows in the output.

The function $F(t)$ is called the antiderivative of $f(t)$ and is found symbolically on the TI in either of two forms. In Figure 15.1 the entry ∫(sin(x),x) gives the antiderivative, -cos(x); while ∫(sin(x),x,c) adds the constant. It is noted that some functions have no analytically defined anti-derivative. When the antiderivative is not known, the output is the original input with as much processing as could be done. In Figure 15.1 the TI does not know an antiderivative of the function fPart(x).

One reason for using this theorem is that it calculates the definite integral in a simple way when we know the antiderivative. For example, we previously used the Riemann sum to guess that the definite integral $\int_0^\pi \sin(x)\,dx = 2$. This is confirmed using an antiderivative

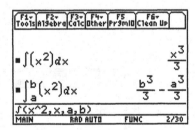

Figure 15.2 The TI uses the Fundamental Theorem to symbolically evaluate a definite integral.

and the Fundamental Theorem. We can even do this calculation in our head: an antiderivative of $f(x) = \sin(x)$ is $F(x) = -\cos(x)$, so

$$\int_0^\pi \sin(x)\,dx = F(\pi) - F(0) = -\cos(\pi) - (-\cos(0)) = -(-1) - (-1) = 2$$

The calculator uses the Fundamental Theorem in its symbolic answers. The first entry shown in Figure 15.2 verifies that the TI knows a symbolic antiderivative of x^2. Then it uses the Fundamental Theorem with $F(x) = x^3/3$ to evaluate the definite integral symbolically as $F(b) - F(a)$.

The definite integral as the total change of an antiderivative

A second reason to use the Fundamental Theorem is that it gives us an exact answer, which may be required or just plain useful. For example, when a growth factor compounds continuously, the decimal accuracy is limited to that of the calculator. This is fine when we are dealing in thousands or millions, but sometimes we have amounts that are astronomical and we want an answer that is correct to whatever number of decimal places is required. Think of the value of π: it is roughly 3, or, if more accuracy is needed, we can use 22/7, or better yet, 3.1459. In its exact form, the symbol π represents full accuracy, not a decimal or fraction approximation.

Let's look at an example where an exact answer is found. Consider a savings account into which you put a dollar every hour. What will it be worth in 20 years if it is compounded continuously at a 10% annual rate? This is a thinly disguised definite integral. First, whatever you deposit needs to be expressed in an annual amount so that all our rates are annual. Call this amount P, which we will take as 365*24 (ignoring leap years). Deposits are so frequent that we consider the rate to be continuous. The future balance in ten years is then given by the definite integral

$$\int_0^{20} Pe^{0.1(20-x)}dx$$

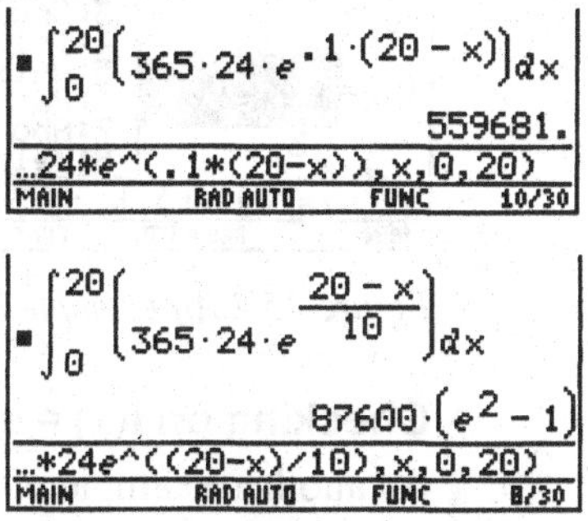

This gives a value of over half a million dollars. This is probably good enough in this case, but the calculator answer does have limited accuracy. Rewriting the equation with no decimals, we see in Figure 15.3 that the Fundamental Theorem is used to write an exact answer (correct to an infinite number of digits). The status line shows that we are in the AUTO mode of evaluation, which works in the EXACT mode (also called *closed form*) whenever possible. When written in terms of e, the expression has full accuracy. Knowing the closed form solution allows us to accurately find the future value of saving a thousand dollars per minute, although we would need a calculator with more internal digits of accuracy to see the difference. The two answers are the same when compared by finding the approximate value of the second integral by using ♦_≈.

Figure 15.3 A definite integral calculated using the closed form given by the Fundamental Theorem. When decimals are involved the approximate answer is shown.

> **Tip:** The MODE setting can be changed from AUTO to EXACT to force all the integrations to be exact. The setting is shown on the status line. It is better to not to change mode settings unless absolutely necessary.

Viewing the Fundamental Theorem graphically

In the above examples, the lower and upper limits were constants. Now, as we did in the last chapter, consider the upper limit as a variable. Specifically, replace b with x on each side of the Fundamental Theorem equation, giving a function in terms of x:

$$\int_a^x f(t)dt = F(x) - F(a)$$

We now consider an example by setting up the following functions:

$f(x) = $ y1 $= $ x^2/10, whose simplest antiderivative is $F(x) = $ y3 $= $ x^3/30

Since a is arbitrary, let $a = -3.5$. (So the integral is zero at $x = -3.5$: there is no width for the area there.) Finally, we define the two functions we want to compare and see if they are equal:

$$y2 = \int(y1(t),t,-3.5,x) \text{ and } y4 = y3(x)-y3(-3.5)$$

Using a `ZoomDec` graphing window (with `xres=2`) and graphing styles y1 (`Thick`), y2 (`Dots`), and y4 (`Path`), you see in Figure 15.4 that y2 and y4 have the same graph.

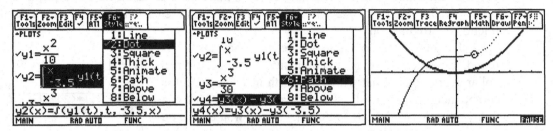

Figure 15.4 An example of the Fundamental Theorem with f(t) = t²/10.

Checking on f(x) = eˣ, the function that is its own antiderivative

As another example of this kind of comparison, we let $f(x)$ be the famous exponential function, whose antiderivative is itself. We see in Figure 15.5 that the general exponential function is its own antiderivative (up to a constant coefficient). This is why the number e is so important: the exponential function with base e is exactly its own derivative and antiderivative.

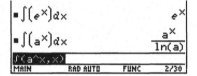

Figure 15.5 The antiderivative of the exponential function.

Comparing d(∫(...)...) and ∫(d(...)...)

What happens when you find the derivative of the integral function? It should not be too surprising that you get the original function back — or do we? In Figure 15.6 we try this out with $y = \sin(x)$ and we confirm that it is true for this function. Disregarding possible differences in constant terms, differentiating and integrating are inverse operations.

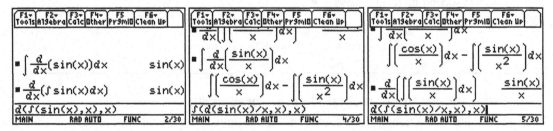

Figure 15.6 Nesting, the derivative of the integral, and vice versa.

But now consider $g(x) = \sin(x)/x$. Even though the TI does not give a symbolic answer to the integral of $g(x)$, it gives the original function for the derivative of the integral. However, this does not work for the integral of the derivative. This example is beyond the calculator's symbolic integration capacity. We can verify that the integral of $g'(x)$ really is $g(x)$ by graphing it. To graph the integral, we use t as the internal variable and x as the upper bound. The choice of the lower bound affects the constant of integration, so the integral is verified if the two graphs are vertical shifts of one another.

Graphing comparison

We are now faced with practical considerations of graphing. It would be clearest to compare
y1=∫(d(sin(t)/t,t),t,1,x) and y2=sin(x)/x, but even with xres=5 the graphing of the
first function is insufferably slow. Entering the formula for $g'(t)$ speeds up the graph as
shown in Figure 15.7. The best option is using the F3 (Calc) option B:nInt instead of ∫; this
speeds up the graphing by about a factor of three. Even so, the graph in Figure 15.7 took
about three minutes. This graph verifies that the integral of the derivative is the original
function, up to a constant.

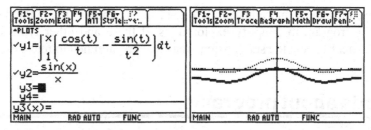

*Figure 15.7 Graph showing that the integral of the derivative
differs from the original function by a vertical shift of sin(1) =
0.84147... The WINDOW setting is ZoomTrig after using ZoomIn
with a factor of 2.*

nInt versus ∫

In Chapter 9, we used the, nDeriv, to give a numeric approximation when confronted with
the rare function that the TI could not differentiate symbolically. The numeric integrator,
nInt, is not usually needed since ∫ uses it for definite integrals without exact answers. See
Figure 15.8

There are many functions that the TI cannot integrate symbolically, for example $f(x) =$
$e^{\wedge}(x^2)$. This is not a deficiency of the calculator, just a fact of calculus — many seemingly
benign indefinite integrals have no closed form solutions. When the calculator is confronted
with a definite integral, it checks to see if it knows the exact formula and, failing that,
switches to a numeric method. Thus, unless you want to save the short time it takes in
checking for an exact formula, you can use ∫ almost all the time. If the MODE is APPROXIMATE
then most exact techniques are ignored. In cases where you know all the work will be done
with numeric approximations, as with the graphing above, use nInt. The numeric methods
for both of these integral commands are similar to the ones discussed in the next chapter.

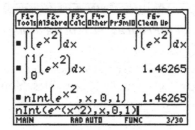

*Figure 15.8 A function without an
exact form integral automatically
uses nInt to calculate the
definite integral.*

CHAPTER SIXTEEN

RIEMANN SUMS

In Chapter 13 we introduced the right- and left-hand sums to approximate the definite integral. In this chapter we use a program to simplify explorations of these and other types of sums. We add the capability to graphically view the subdivision areas that are summed to make the approximation.

A few words about programs

A program is a set of commands that are performed in a prescribed order, like a recipe. The order is normally the sequential list of commands, but there are techniques to alter that order. Special program command menus (I/O and CTL) list the input/output and control commands. A program is written by pressing APPS 7:Program Editor 3:New... and arrowing down to enter a program name in the Variable: box. The first choice (Type:) in the dialog box allows you to choose between a program and function, the second choice (Folder:) lets you assign a folder to work from. In this book, all work is done in the default folder named MAIN. If you want edit the last program you have used, then press APPS 7:Program Editor 1:Current and the current program is put on the screen for editing. A colon automatically begins each new program line. Use 2nd_QUIT to exit the Program Editor. If your MODE setting for Apps Desktop is ON then Program Editor appears as an icon, not as a list as it appears in Figure 16.1.

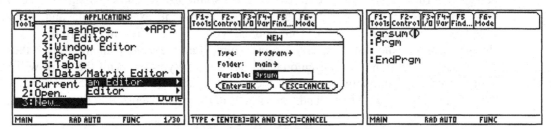

Figure 16.1 The APPS menu 7:Program Editor. A dialog box allows you to name a new program.

Tip: Use program names of at least two characters so that they are not deleted when you periodically clear the single letter variables. You cannot store a number in a variable that has been named as a program.

Tip: To edit the current program, the key sequence APPS 7 1 allows quick entry to the listing.

Entering a program from the printed listing below is quite tedious and you can expect to make a few errors that are only discovered when executing the program. However, once a program is correctly entered, it can be transferred to other TI calculators of this class using the built-in Link features. In the classroom, it is typical that a program is verified by the instructor and distributed to the class using 2nd_VAR-LINK F3 (Link) 1:Send. If you downloaded the Calculus Tools as an APP then that application gives you much the same capability as these two programs. Enter the program before using the next section.

Using programs to find and view Riemann sums

A program is activated — the more common terms are *run* or *executed* — by typing its name with parentheses on the entry line and pressing ENTER. For example rsum(). This program is designed to require a minimum of input within the actual program. Before using it, you must define your function in y1 and set the window to have xmin be the left endpoint and xmax be the right endpoint of the desired interval.

Program rsum()

Figure 16.2 shows the window and graph of y1=sin(x^2). On the home screen, enter the program name rsum(). The program prompts for the number of partitions (subdivisions) of the interval. Five different kinds of Riemann sums are calculated and displayed. The value labeled nInt is short for nInt(y1,x,xmin,xmax) and is used to judge the other numerical approximations. The last value, Simpson, is the value given by is a weighted average of the two previous results, specifically the sum of the trapezoid value and twice the midpoint value all divided by three. This value is consistently close to the 'true' value for small partitions. The nInt value comes from a more complex weighted average.

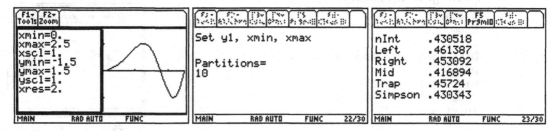

Figure 16.2 Set y1=sin(x^2) and the window shown before using the rsum() on the entry line of the home screen.

> *Tip:* When you exit a program, use HOME to return to the home screen.

Program grsum()

In Figure 16.3, the graphic representation of a Riemann sums is shown in the order of the four menu choices on the menu line. The function setup and window is the same as Figure 16.2. The number of partitions is 10; this is internally reset if needed. Press HOME when you have finished viewing the graph. Press ENTER and the program starts again with the four menu choices.

The two programs listed below take a considerable amount of time to enter, but once entered, they can be passed to other TI calculators of similar type. (They are in the public domain.) They can also be stored on a computer and passed to other calculators using TI-Connect.

If you are entering these by hand, use a copy-and-paste technique (♦_COPY & ♦_PASTE) on repetitive code. Special symbols such as Σ are found in the CATALOG. Once you have this program working and if you are feeling adventurous, you might modify it to use more complex TI programming features, like a toolbar for pull down menus.

Program rsum() listing

```
rsum()
Prgm
ClrIO
Disp "Set y1, xmin, xmax"
Disp ""
Input"Partitions= ",n
(xmax-xmin)/n→d
ClrIO
Output 1,1,"nInt"
Output 1,50,nInt(y1(x),x,xmin,xmax)
Output 11,1,"Left"
Σ(y1(xmin+i*d),i,0,n-1,1))*d→s1
Output 11,50,s1
Output 21,1,"Right"
Σ(y1(xmin+i*d),i,1,n,1))*d→s2
Output 21,50,s2
Output 31,1,"Mid"
Σ(y1(xmin+i*d),i,.5,n,1))*d→s3
Output 31,50,s3
Output 41,1,"Trap"
(s1+s2)/2→s4
Output 41,50,s4
Output 51,1,"Simpson"
Output 51,50,(2*s3+s4)/3
EndPrgm
```

Program grsum() listing

```
grsum()
Prgm
ClrIO
Disp "Set y1, xmin, xmax"
Disp ""
Disp "1=Left    2=Right"
Disp "3=Mid     4=Trap"
Input a
ClrIO
ClrDraw
10→n
(xmax-xmin)/n→d
For i,0,n-1
xmin+I*d→x
If a=1 Then
y1(x)→ya:ya→yb
EndIf
If a=2 Then
y1(x+d)→ya:ya→yb
EndIf
If a=3 Then
y1(x+d/2)→ya:ya→yb
EndIf
If a=4 Then
y1(x)→ya:y1(x+d)→yb
EndIf
Line x,0,x,ya
Line x,ya,x+d,yb
Linex+d,yb,x+d,0
EndFor
EndPrgm
```

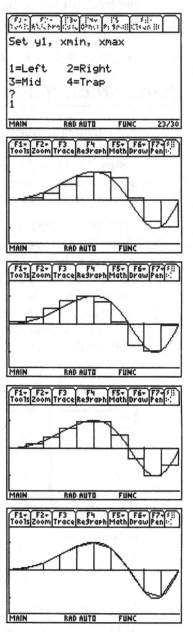

*Figure 16.3 Using the grsum()
in the order of the menu.*

CHAPTER SEVENTEEN

IMPROPER INTEGRALS

In this chapter we look at two different problems encountered when using the integral in a wider setting. These special cases involve infinity and are called improper integrals. First, we see that the limit of integration can be infinite. Second, we see that integration is sometimes possible even when the integrand function itself has infinite (undefined) values.

An infinite limit of integration

On first take, you might think that any positive function that goes on forever must have an infinite definite integral. Let's try a thought experiment. Suppose you decided to go on a diet and every day you cut your chocolate chip cookie consumption in half. How many cookies would you need for your lifetime? (or for eternity?) Visualize the cookie: the first day you would eat half, the next day a half of a half (a quarter), and so on. Because each day you would only eat half of the remaining cookie, you would never finish it: one cookie would last a lifetime! Figure 17.1 shows the sum is 1. (The ♦_∞ symbol is above CATALOG.)

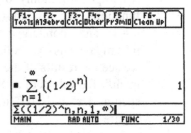

Figure 17.1 The sum of cookie halving is one.

Three power functions

Let's compare the three functions

$$y_1 = \frac{1}{x}, \; y_2 = \frac{1}{x^3} \text{ and } y_3 = \frac{1}{x^{1/3}}$$

as x goes from 1 to infinity and see if we get a graphical hint about which might have a finite area under it. For this to happen, the function must approach zero. We see in Figure 17.2 that all three functions do approach zero as x gets large, and y2 does so the fastest. Just like the sum in Figure 17.1, we can find the improper integral by using infinity as the upper limit.

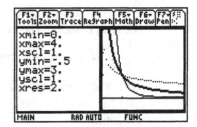

Figure 17.2 Comparing three functions as x goes to infinity. Styles are Thick for y1, Line for y2, and Dot for y3.

In Figure 17.3 we see that the integral for y2 converges, but not those for y1 and y3. However, these divergent integrals are assigned finite values in the APPROX mode. This is because the calculator's numeric method erroneously decided at some point that it could ignore the decreasing terms in its area summation. Pay attention to the evaluation mode and question results that are done in the APPROX mode. If your functions involve decimals, then you need to change the MODE setting from AUTO to EXACT to be sure of answers to improper integrals.

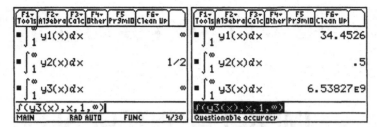

Figure 17.3 Evaluating improper integrals with EXACT and APPROX settings, the latter giving incorrect answers.

Tip: When evaluating improper integrals, expect long calculation times. There is no way good way to speed this up. Should the wait be too long to bear, press ON to halt the computation.

The story on $\int_1^\infty \frac{1}{x^p}dx$

An important family of functions is the negative power functions. We saw above that the integral diverged when $p = 1$ and converged when $p = 3$. We can further investigate this symbolically with the TI. In Figure 17.4, we see the results of the general improper integral when we qualify the values of p. You can investigate other examples such as $1/x^2$ and $1/x^{1/4}$ to test the analytic results shown in Figure 17.4.

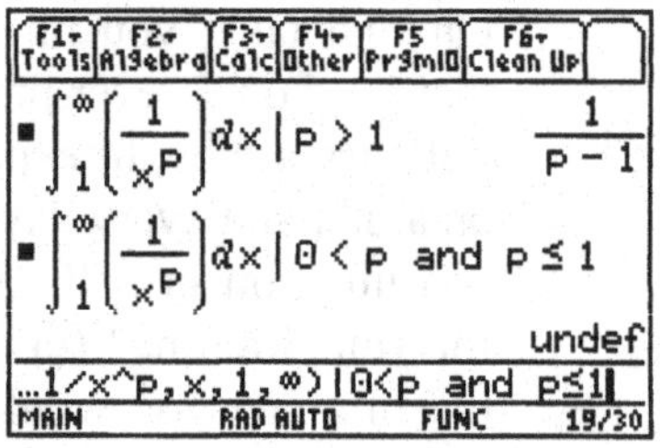

Figure 17.4 The negative power function converges when $p > 1$.

The convergence of $\int_0^\infty \frac{1}{e^{ax}}dx, a > 0$

It is simple to show analytically that $\int_0^\infty \frac{1}{e^{ax}}dx = \frac{1}{a}$ for $a > 0$. Using the TI to find this symbolic answer, we must avoid two common errors. The first line of Figure 17.5 does not give us what we expect because of a common entry error: no multiplication sign was inserted between a and x. The calculator understood ax as a constant and gave an answer based upon $\frac{1}{e^{ax}}\int_0^\infty dx$. With that corrected, we next find a numeric answer of 1/4 when we expected a symbolic one. The variable a has a previously defined value of 4, and must be cleared (DelVar a) to be used as a symbolic variable. Finally, the last entry gives the result we are expecting.

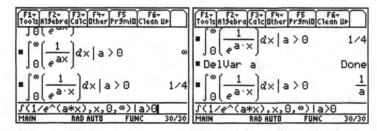

Figure 17.5 Symbolic evaluation of an improper integral beginning with two mistakes.

The integrand goes infinite

The second type of improper integral is an infinite integrand (one version of being undefined). This is a much more dangerous situation because there is no ∞ symbol in the integral to alert you. It is a good habit to graph a function before finding the definite integral. The graph alerts you to potential problems, like the integrand being undefined (and tending to positive or negative infinity). In Figure 17.6, we first blindly find the definite integral with the value of the integrand being infinite at $x = 4$. By drawing a graph, we can see why the integral diverged, but finding the integral from the graph gives us a numeric answer (but warns us about questionable accuracy). The second example shows an improper integral that converges even though the interval includes a point where the integrand goes to infinity.

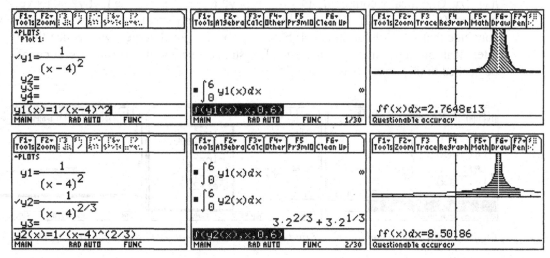

Figure 17.6 An improper integral with finite limits that diverges and an infinite integrand that converges.

Results that make us wonder

We have seen that the integral function may give misleading (incorrect) results when using APPROX (either with the MODE setting or ♦_≈). Another example, in Figure 17.7, shows an attempt to integrate over an interval not in the function's domain. In one case, the calculator gives an answer which we may not recognize as non-real. In the second (very similar) case, we are alerted by an error message that the answer is non-real.

A calculus teacher can easily make up examples where the calculator misleads you. Would an instructor really do that?

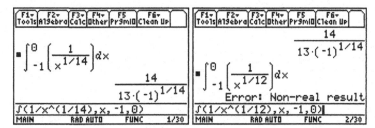

Figure 17.7 Strange answers for integrals over intervals outside the functions' domain.

The comparison test

The negative power function and the negative exponential function are of special importance because they are often used as a benchmark for comparison with more complicated functions. For example, suppose we wanted to know if the integral

$$\int_4^\infty \frac{1}{\ln x - 1}\, dx$$

converges or diverges. In Figure 17.8, the symbolic answer of ∞ indicates we should suspect divergence. Evaluating in APPROX mode gives us, a very large number as an answer, and the status line warning questionable accuracy. To be sure we can use the comparison test with $f(x)=1/x$. Graphically comparing this function y2 to y1=1/x, we see that y2 is above y1 in the window ($4 \le x \le 1000$). Knowing the long-term behavior of the natural logarithm we can conclude that y2 > y1 for all $x \ge 4$. Since $\int(1/x,x,4,\infty)$ diverges, we deduce that the larger integral in question must also diverge. The status warning questionable accuracy should be heeded in this case.

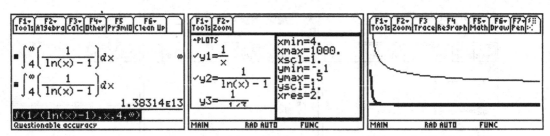

Figure 17.8 A shaky symbolic result and a graph showing that y2 (Line) exceeds y1 (Thick).

Tip: Watch the status line when doing integration. The warning questionable accuracy should put you on guard, especially when answers are near 1E13.

CHAPTER EIGHTEEN

APPLICATIONS OF THE INTEGRAL

We look at applications of the integral. These typical examples give only a flavor of the extensive applications of the integral.

Geometry: arc length

The following calculus formula finds arc length along a function's graph.

$$\text{Arc length} = L = \int_a^b \sqrt{1 + (f'(x))^2}\, dx$$

Let's calculate the arc length of the curve $y = x^3$ from $x = 0$ to $x = 5$ and compare it to direct distance from the origin to the point (5, 125). In Figure 18.1, we see this result calculated on the home screen in two ways. It should be immediately obvious that the built in arcLen command is more efficient to use. To correctly enter the syntax of the integral expression is a challange.

From a graph screen the TI calculates both arc length and distance between points. Use the F5 (Math) menu and select B:Arc or 9:Distance. The usual bound prompts appear, this time as 1st Point? and 2nd Point?. Both results are shown in Figure 18.2. Notice that a direct distance line is drawn on the graph as part of the Distance computation. The graph's scale makes the difference between the two lengths look greater than the actual numerical difference.

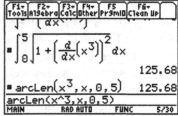

Figure 18.1 Arc length found in two different ways.

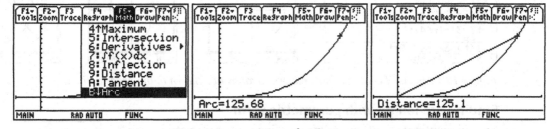

Figure 18.2 Finding the arc length of the curve $y = x^3$ from $x = 0$ to $x = 5$ and comparing it to the direct distance between (0, 0) and (5, 125). Window: $-1 \le x \le 6$ and $-25 \le y \le 150$.

Parametric graphing

To graph parametric equations change Graph option in MODE to PARAMETRIC. (PAR is shown on the status line) In this kind of graph, x and y are independently defined in terms of a variable t (usually thought of as time). As a parametrically defined curve, a circle is composed of the sine and cosine functions. In Figure 18.3 the MODE is set to PARAMETRIC and the equation of a circle with radius 3 is defined. Notice the changes in the Y= editor: equations are now in pairs and the independent variable is t. We set a ZoomDec window and trace is used to see values around the circle for $0 \le t \le 2\pi$.

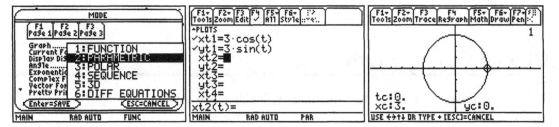

Figure 18.3 Select PARAMETRIC, define a pair of parametric equations, and graph in a ZoomDec window.

> **Tip:**   If the t values of Figure 18.3 are not $0 \le t \le 2\pi$, they can be quickly reset using ZoomStd followed by ZoomDec. The only Zoom feature that changes the t-settings is ZoomStd. All others Zoom options reset the window size, but not the t-values.

Changing parameters

In Figure 18.4 we look at the effect of changing the parameters of the equations defined in Figure 18.3. The first curve appears the same, but is an incomplete circle since the sine and cosine values only went nine-tenths of the way from 0 to 2π. If you increase the coefficient of t to 3: the curve is wrapped around the same path three times. In the other screens, a lower coefficient on the cosine creates an ellipse and a mixture of periods results in a beautiful pattern.

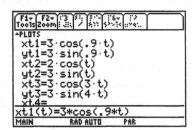 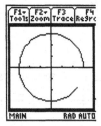 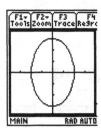 

Figure 18.4 Effects of changing parameters. Graphs are shown in the order defined.

Finding arc length with parametric form

The circumference of an ellipse defined in parametric equations is given by the formula:

$$\text{Arc length} = L = \int_a^b \sqrt{\left(\tfrac{dx}{dt}\right)^2 + \left(\tfrac{dy}{dt}\right)^2}\, dt .$$

For example, the circumference of the parabola shown in Figure 18.4 is calculated in Figure 18.5.

$$\blacksquare \int_0^{2 \cdot \pi} \sqrt{(xt1(t))^2 + (yt1(t))^2}\, d\blacktriangleright$$

$$15.8654$$

...1(t)^2+yt1(t)^2),t,0,2π)
MAIN   RAD AUTO   PAR   1/30

Figure 18.5 Arc length of a parabola from its parametric definition.

Writing y = f (x) functions in parametric form

Any function $y = f(x)$ can be written in parametric form by setting xt1=t and yt1 to $f(t)$. We see in Figure 18.6 that our first graph is restricted to the sine curve with $0 \le t \le 2\pi$ (the ZoomTrig setting). In the next graph, we changed t so that the graph extends across the origin.

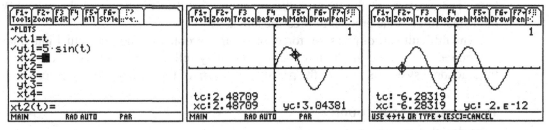

Figure 18.6 Graphing y = f(t)=5sin(t) parametrically in ZoomStd. Reset tmin=-2π for more graph.

Polar coordinate geometry

To graph using polar functions and coordinates change the Graph option in MODE from Function to Polar. The Y= editor screen now lists the functions as r1 to r99, and POL appears on the status line. In Figure 18.7, we graph r1=3sin(2θ) and find the arc length of one petal, $0 \le \theta \le \pi/2$.

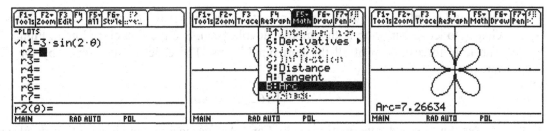

Figure 18.7 Arc length found for one petal of a rose in Polar coordinate mode. (1st Point? =0 and 2nd Point?. = π/2.)

> ***Tip:*** Do not use arcLen from the home screen to find arc length on a polar function. It gives incorrect values.

The MODE must be Polar to graph polar equations, but if you do not need a graph, or after you have seen a graph, you can return to Function mode and evaluate polar formulas from the home screen entry line. The area of an enclosed region in polar coordinates is

$$\text{Area of enclosed region} = \tfrac{1}{2} \int_{\alpha}^{\beta} f(\theta)^2 \, d\theta$$

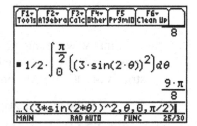

Figure 18.8 Area of one petal. Same value in Func or Pol mode.

The area of one rose petal is calculated in Figure 18.8. The same result is given regardless of whether the mode is Polar or Function.

Physics: force and pressure

Water pressure increases with depth so that there is more pressure at the bottom of a container than at the top. One cubic foot of water weighs 62.4 pounds and exerts a force of 62.4 pounds on the base of its container. With half a cubic foot, the force is 62.4/2 = 31.2 pounds on the base. The pressure on the base is directly proportional to the depth of the water. We also know that force is the product of pressure and area.

The difficult part of pressure, force, or volume problems is not the actual integration that is required, but setting up the integral to accurately reflect the geometry of the situation. A time-honored system is to write the pressure as a sum of forces acting on strips or slices.

The pressure on a trough

Consider the trough shown in Figure 18.9. There are four sides that the force of water acts on. Let's tackle the easiest side first: the 3' by 14' horizontal back. We subdivide the height into pieces of length Δh, so this back side is made up of horizontal strips, each having an area of $14 \cdot \Delta h$ square feet. The entirety of each strip is at the same depth, so that all along a given strip there is equal pressure, namely

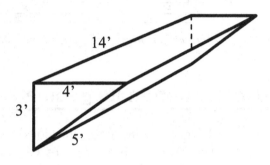

Figure 18.9 A trough with the shape of a triangular prism.

force on a horizontal strip = $(62.4h)(14 \cdot \Delta h)$

We could simplify by multiplying constants together, but the calculator can do that and we choose to leave our setup in this more readable form. See the first entry line in Figure 18.10. The total force is sum of all the horizontal strips. The exact total is given by the definite integral

$$\int_0^3 (62.4h)(14)dh$$

Next we consider the inclined side. For a Δh (vertical height change) the actual width w on the inclined side is greater than Δh. Using similar triangles, these values are related by $\Delta h/w = 3/5$. The area of the strip is $14(5/3)\Delta h$, so

force on inclined strip = $(62.4h)(14)(5/3)\Delta h$

The total force is given by the integral

$$\int_0^3 (62.4h)(14)(\tfrac{5}{3})dh$$

Finally we compute the force on the two triangular ends. Again using similar triangles, we find the area of a strip is $(4/3)(3-h)\Delta h$ and

force on end strip = $(62.4h)(4/3)(3-h)\Delta h$

so that the third integral is

$$\int_0^3 (62.4h)(\tfrac{4}{3})(3-h)dh$$

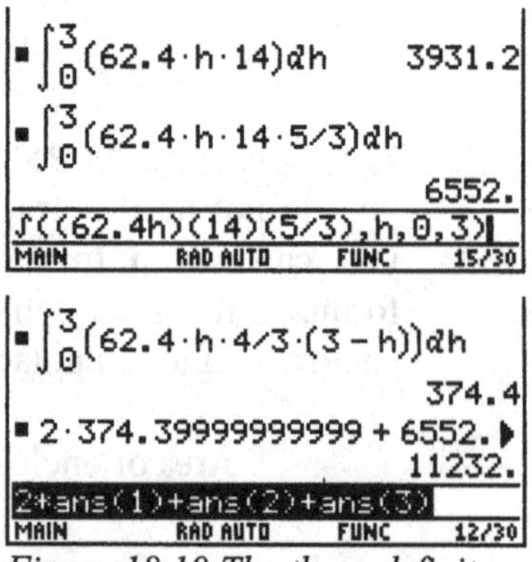

Figure 18.10 The three definite integrals to find the force on a trough.

Now the easy part is entering these definite integrals for evaluation and computing the final sum; this is done in the second frame of Figure 18.10 as shown highlighted on the entry line.

Tip: It is much safer to enter expressions in an expanded form so that the derivation remains evident.

Economics: present and future value

Suppose you win a two million dollar lottery. Before you spend the money, you are told that the money is distributed to you over the next twenty years. That is a mere $100,000 per year (before taxes). If you wanted to get an immediate lump sum, you could sell your rights to all the future payments. What is this really worth? In economics this value is called the present value, V. It is calculated using a fixed investment rate and an integral. If $P(t)$ is the annual income stream and r is the annual investment rate for M years, then the present value is

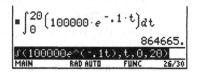

Figure 18.11 The present value of a two million dollar lottery. Payments are $100,000 per year for 20 years.

$$V = \int_0^M P(t)e^{-rt}\,dt$$

Suppose that the agreed investment rate was 10% for the twenty years. Then we enter the integral as shown in Figure 18.11. The present value is $864,665, less than half the advertised amount!

Using Numeric Solver with nInt

In the APPS menu A:Numeric Solver allows us to both evaluate the present value and see the effect of different investment rates. We use nInt rather than ∫, since it is better suited for looking solely at numeric values.

In Figure 18.12, after entering the equation, we see the variables listed that were in the equation. We check our setup by recalculating the present value of twenty annual payments of $100,000 with an investment rate of 10%. To do this, enter p=100000, r=.1, and m=20. Move to the v line, press F2 (Solve), and we get the same answer as in Figure 18.11. The left-rt value that appears is an indication of the answer's accuracy. Since t is a dummy variable of integration, it does not appear in the variable list.

Next we solve for an investment rate given a present value of $500,000. Remember that recalculation takes place when you place the cursor on the desired variable's line and press F2 (Solve). We find a present value of half a million corresponds to an investment rate of about 20%; in Figure 18.12, $r = .196\ldots$. Use HOME to exit.

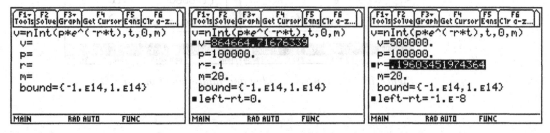

Figure 18.12 Numeric Solver finding the present value v and then an investment rate r.

Discrete vs. continuous compounding

The analysis above makes the assumption that the income is coming in continuously, effectively every second, which is not the case. In real life, the money comes in discrete payments: the first is now ($t = 0$), the second in a year ($t = 1$), and so on until the twentieth payment ($t = 19$). If we want to calculate this to the penny, we need to use a discrete sum for the twenty years. The point is that there is a close relationship between the integral and the sum of a payment sequence. The integral must be used when the income is continuous, but it is also commonly used as an approximation for a discrete sum. The sum, of about $900,000 shown in Figure 18.13, is more than the integral calculation because you receive all $100,000 payments at the start of the year. Think of the continuous model as being paid about three-tenths of a cent per second for the next twenty years.

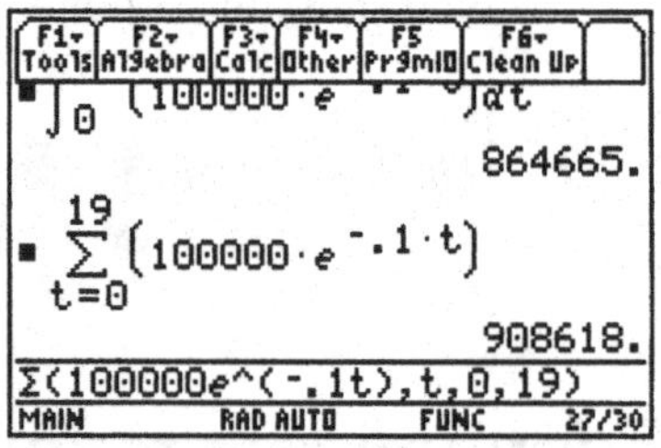

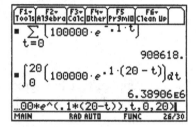

Figure 18.13 The present value is greater when discrete payments are made at the start of each year.

The future value

Since the probability of winning the lottery is essentially zero, you might want to create a jackpot for yourself by investing $100,000 a year (and remember, that's just three-tenths of a cent per second). Your total after M years is called the future value and is given by the definite integral

$$\text{Future value} = \int_0^M P(t)e^{r(M-t)}\,dt$$

Using this formula in Figure 18.14, after twenty years you would have a real jackpot worth over six million dollars.

Figure 18.14 The future value in 20 years.

Sometimes the exact forms of an integration show hidden relationships. For this exercise we in Figure 18.15, we start by clearing the variables to insure symbolic results. Change the decimals to fractions in the equations, and recompute the integrals for present and future value. We find that

$$\text{Present value} = \text{future value} \cdot (e^{-mr}).$$

This is an example of the usefulness of the Fundamental Theorem: you would not see the e^{-mr} factor without writing the integrals in symbolic form.

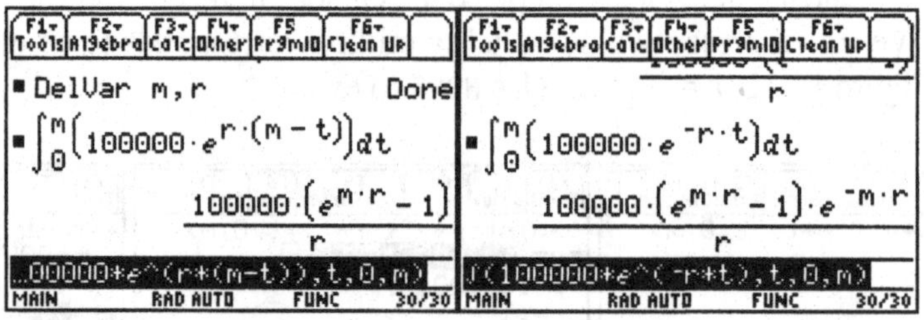

Figure 18.15 The present value and future value in exact form.

Modeling: normal distributions

In statistics, a normal distribution has a graph that is a bell-shaped curve. Its general equation is

$$p(x, \mu, \sigma) = \frac{1}{\sigma\sqrt{2\pi}} e^{-(x-\mu)^2/(2\sigma^2)}$$

where μ is the mean and σ is the standard deviation. We use a two letter name so it remains defined even when we clear variables. Define it on the home screen by

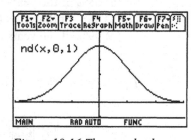

Figure 18.16 The standard normal curve.

```
Define nd(x,μ,σ)=1/(σ*√(2π))e^(-(x-μ)^2/(2σ^2))
```

The Greek letters are entered by using the 2nd_CHAR, Greek. We then use this to graph the function with specific values of $\mu = 0$ and $\sigma = 1$. The resulting curve is called the *standard* normal curve and is shown in Figure 18.16 with the window $-3 \leq x \leq 3$ and $-.1 \leq y \leq .5$. We saw an example of this family in the last section of Chapter 12.

The mean and standard deviation

Two fundamental concepts from statistics are the mean of a set of values (casually called the average) and the standard deviation, a measure of how spread-out the values are. The commands mean and stdDev, available in the CATALOG, calculate these values for any list. If a list is previously stored and you are not sure of its name, then you can paste it from the 2nd_VAR-LINK menu. For example, in Figure 18.17, we press H to find the variable height that we used in a previous chapter. Next we find the mean and standard deviation of this list.

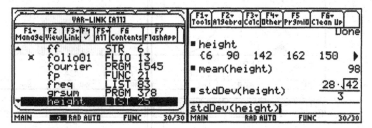

Figure 18.17 Finding the mean and standard deviation for values of a list.

The Anchorage annual rainfall

One application of the normal distribution is to model situations where measurements are taken under conditions of randomness. For example, suppose you look at the records for annual rainfall in Anchorage, Alaska over the past 100 years. Let's simplify and say that you found the average of these averages to be 15 inches. Let's say that the standard deviation was 1. We can estimate the fraction of the years that rainfall is between

    (a) 14 and 16 inches,

    (b) 13 and 17 inches, and

    (c) 12 and 18 inches.

Taking three integrals of the normal distribution with μ = 15 and σ = 1. Using F5 (Calc), we see from the graphs in Figure 18.18 that the model predicts
(a) 68% of the years have rainfall between 14 and 16 inches,
(b) 95% of the years have rainfall between 13 and 17 inches, and
(c) 99% of the years have rainfall between 12 and 18 inches.

Tip: After graphing an integral, use F4 (ReGraph) to clear the previous drawing.

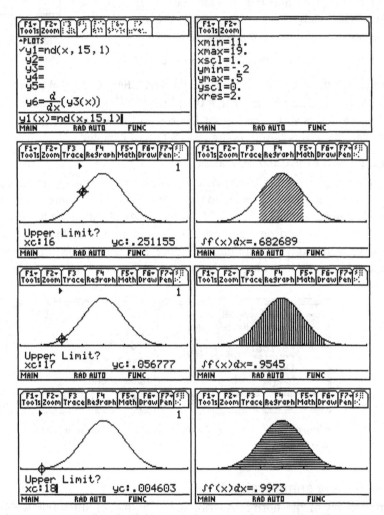

Figure 18.18 Graph of the normal curve with mean 15 and standard deviation 1, then using F5 menu to find the integrals with limits (a) 14 ≤ x ≤ 16, (b) 13 ≤ x ≤ 17 and (c) 12 ≤ x ≤ 18. These values confirm the famous 68-95-99 rule of normal distributions.

A note about TI statistical features

We use statistics again in Chapter 24, but the TI has powerful statistical features that are not addressed in this book. See the *TI Guidebook*.

PART IV SERIES

CHAPTER NINETEEN

GEOMETRIC AND CONVERGING SERIES

A series is the sum of a sequence of numbers. We look at finding a function from a given series and in some cases, the TI writes a general formula for a sum. The calculator helps calculate finite sums, find general formulas and, with some success, determine whether a series converges or diverges.

The general formula for a finite geometric series

A finite geometric series has the form

$$a + ax + ax^2 + \ldots + ax^{n-1} + ax^n$$

Note that the coefficient a is the same for each term. The closed form sum is

$$S_n = \sum_{i=1}^{n} ax^{i-1} = a\frac{(1-x^n)}{(1-x)}, \quad (x \neq 1)$$

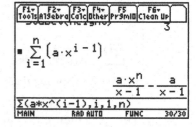

Figure 19.1 The finite geometric series sum.

With the input for this finite sum formula shown on the entry line in Figure 19.1, the output is equivalent to the traditional form. You might think that such a sum would be so rare as to make it not worth considering, but this kind of finite series comes up in many situations.

Repeated drug dosage using

Consider a 250 milligram dose of an antibiotic taken every six hours for many days. The body retains only 4% of the drug present in the body after six hours. Be careful: this does not say that 4% of 250 mg is left. This is only the case at the end of the first six hours. The interesting part is that at the end of the second six hours, the body retains 4% of the second dose *and* 4% of the remaining first dose. Let's make a sequence of the amount of drug in the body right after taking the n-th dose.

$$Q_1 = 250,$$

$$Q_2 = Q_1(0.04) + 250,$$

$$Q_3 = Q_2(0.04) + 250, \text{ etc.}$$

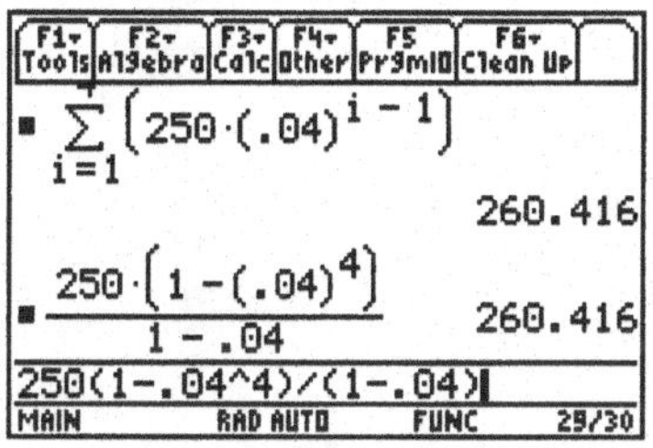

But if we substitute lower sums and multiply, we find

$$Q_2 = 250(0.04) + 250 \text{ and}$$

Figure 19.2 The dosage present in the body at the fourth dose summed as a sequence and compared to the formula.

$$Q_3 = 250(0.04)^2 + 250(0.04) + 250$$

In general,

$$Q_n = 250(0.04)^{n-1} + \ldots + 250(0.04)^2 + 250(0.04) + 250$$

The Q's are a finite geometric series with $a = 250$ and $x = 0.04$. Let's calculate the amount of antibiotic in the body at the time of taking the fourth pill. In Figure 19.2 we use Σ to find the total amount left after the fourth pill. This is the same answer we get from the series formula.

Using a table to find consecutive finite sums of a sequence

What happens as n increases? To see consecutive finite sums of a sequence, you can define a function using Σ and having a variable upper limit. In Figure 19.3 we set a function in y1 and set up a table to show the desired values. Clearly, the table settings start at 1 and have increment 1 to insure the x-values are positive integers. The middle frame of Figure 20.3 shows the use of use ♦_| (Format) to increase the cell width to 12. Notice how the amount of medication is quite stabilizes by the end of the first day (by the fifth dose).

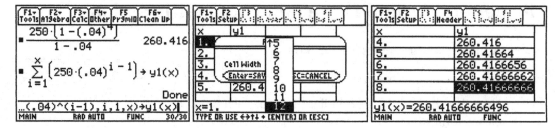

Figure 19.3 Showing finite sums using a function and Table.

> *Tip:* In Figure 19.3 the Display Digits mode has been set to FLOAT in order to see as much accuracy as possible.

A different drug might have a much higher retention level. Redefining y1 from inside the table with 0.50 instead of 0.04 gives the table in Figure 19.4. A drug with this retention level takes about two days (10 doses) to reach its stable limit, near 500 mg.

> *Tip:* A column formula is change by having the cursor in the desired column and pressing F4 Header.

x	y1
10.	499.51171875
11.	499.75585938
12.	499.87792969
13.	499.93896484
14.	499.96948242

y1(x)=Σ(250*(.5)^(i-1),i,...
MAIN   RAD AUTO   FUNC

Figure 19.4 Retention amounts for the 50% rate.

Regular deposits to a savings account

Another direct application of the finite geometric series is finding the value of an investment that earns a periodic interest. Suppose that you plan for retirement by putting $1000 a year into a savings account that earns 5% annual interest. You want to know how much this is worth after the nth deposit. We find this in Figure 19.5 by entering the summation into y1 and making a table showing the value every ten years after the first year. Check the sum $n = 41$, is that enough for retirement?

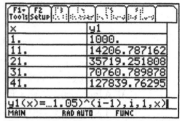

Figure 19.5 The values of a 5% savings account after annual deposits of $1000.

Identifying the parameters of an infinite geometric series

Evaluating a finite series yields a numeric sum, but infinite series may converge or diverge. If the numeric series is geometric, then finding the sum is quite easily. We practice on the following list of infinite series.

$$\text{(a) } 1 + \frac{1}{2} + \frac{1}{4} + \frac{1}{8} + \cdots$$

$$\text{(b) } 1 + 2 + 4 + 8 + \cdots$$

$$\text{(c) } 6 - 2 + \frac{2}{3} - \frac{2}{9} + \frac{2}{27} - \cdots$$

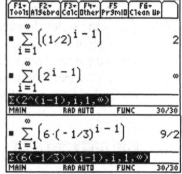

These three geometric sums are all of the form $a + ax + ax^2 + \cdots + ax^n \cdots$, so we need to identify a and x. This is the hardest part. If the sum is given in sigma sum notation, then this task is essentially done for us. In the above cases, we see that each series is a geometric series with a as the first term and we find x from the second term.

Figure 19.6 Checking 3 different series for convergence.

The TI evaluates the sums as shown in Figure 19.6: (a) converges to 2, (b) diverges, and (c) converges to 4.5.

Summing an infinite geometric series by the formula

It is not surprising that if the value of x is one or greater, then an infinite geometric series diverges. If we restrict x to $0 < x < 1$, then the series has a finite sum such the above sum (a) with $x = 1/2$. In Figure 19.7 we see that a general infinite series is not evaluated without restrictions. But the TI does not evaluate the more general formula that allows x to be

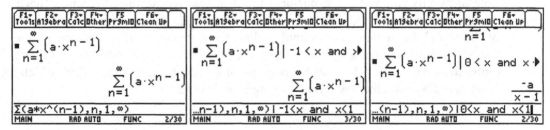

Figure 19.7 The first two frames shows no result, but the restriction $0<x<1$ gives a general sum formula for the infinite geometric series.

negative. The more general formula is

$$a + ax + ax^2 + \ldots + ax^{n-1} + ax^n + \ldots = \frac{a}{1-x}, \quad \text{for } |x| < 1$$

As an exercise, use this formula to confirm the values of our previous two sums (a) and (c). Also, revisit the formula for the two drug dosage cases, they both satisfy the condition $|x| < 1$ since $x = 0.04$ and $x = 0.5$.

Piggy-bank vs. trust

Suppose that your parents are trying to decide on a plan to provide for your future and they have two choices:

I. Each year they put your age in dollars into a piggy bank.

II. Each year they put $3 into a trust account that earns 6% annual interest.

We already know how plan II works: use $a = 3$ and $x = 1.06$ in the formula for a finite geometric series. In plan I we need to look at the sum of the series $1 + 2 + 3 + \ldots + n$. This is *not* a geometric. Comparing table values shows that the piggy bank is probably the better plan for you now and for a while.

To compare the two plans, make a table as shown in Figure 19.8. We see that plan II is preferable from age 11 to 61 and, but as you get into your seventies, plan II is much better.

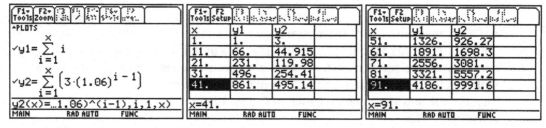

Figure 19.8 Sigma summations can be used to define functions. The table values show y2 exceeds y1 after 61 and before 71.

How can we know if a series converges?

This is a difficult question, but we have the ratio test and the alternating series test to help us in many cases. If asked for a proof, you need to use an analytical argument, but a graph or table often tells you whether to pursue a proof of convergence or divergence. Since you are investigating a series, the n-th term is given symbolically (or else you need to write it symbolically). In the harmonic example below, the sigma summation notation is given, but if a series is listed without a symbolic term, then the first order of business is to write it as a sigma sum. We use y1 = Σ (...) to either graph or make a table of values from which we form an opinion about whether the series converges or diverges.

The harmonic series: $1 + \dfrac{1}{2} + \dfrac{1}{3} + \dfrac{1}{4} + \ldots = \displaystyle\sum_{n=1}^{\infty} \dfrac{1}{n}$

In Figure 19.9 we enter this infinite series as

$$\Sigma(1/n,n,1,\infty)$$

and find that we get no information from this approach. Next we translate the partial sums of the harmonic series into a function that sums the first through x-th terms,

$$y1 = \Sigma(1/n,n,1,x)$$

and graph with `xmin=0` and `xmax=300`. The graphing gets progressively slower as x gets larger Remember that you can press `ON` to interrupt at any time. The partial sums appear to be continually increasing without limit. This means the series does not converge, but analytic means are needed to be absolutely sure.

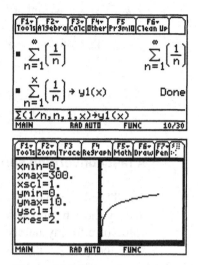

Figure 19.9 Suspecting divergence from a partial sum graph.

You might ask why we don't just find the 10,000th partial sum from the `HOME` screen since the limit value would likely be clear by then. This can be done and the answer is just over 12, but you pay dearly in time and battery use. This calculation takes about one hour on a TI, depending upon the model. Also, without other values for comparison, it is unclear from just this one calculation whether the series is converging or diverging.

The alternating harmonic series: $1 - \dfrac{1}{2} + \dfrac{1}{3} - \dfrac{1}{4} + \ldots = \displaystyle\sum_{n=1}^{\infty} (-1)^{n-1} \cdot \dfrac{1}{n}$

Often series alternate positive and negative terms. This sign switch is cleverly written in sigma notation using a power of negative one. Since every other term is subtracted, we expect the range of the partial sums to not be as great as for the harmonic series. Any alternating series converges if its terms approach zero as n becomes infinite.

When we enter an infinite series on the TI to find a finite sum, we may be greeted with a transformation of the series and no limiting value. The conversion to $\cos(n{\cdot}\pi)$ in Figure 19.10 is yet another clever way of writing a quantity that will alternate between 1 and -1.

We define the partial sums function

$$y1 = \Sigma((-1)^{\wedge}(n-1)*(1/n),n,1,x)$$

and graph it. Press `ON` to break, when you are satisfied. We see from the graph in Figure 19.10 that it is a good bet that this series converges. But unless we waited for the complete graph and used `Trace`, we would not have a very accurate convergence value from the graph. Instead, we could use a table to find a good limiting value.

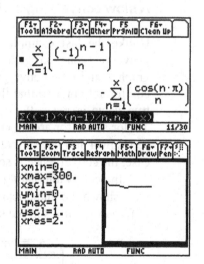

Figure 19.10 No result is given from the HOME screen, but graphing shows a suspected convergence. We break before the graph is completed.

> *Tip:*   When writing a sigma sum symbolically, it is easy to make small errors. Before
> graphing or making a table, you might want to make a small list of terms to verify
> that your notation is correct.

> *Tip:*   If you do not want a sum to be a rational number answer, enter the start or stop value
> as a decimal to put the answer in the APPROXIMATE format.

A fast converging series: $1 + \dfrac{1}{2!} + \dfrac{1}{3!} + \dfrac{1}{4!} + \ldots = \displaystyle\sum_{n=1}^{\infty} \dfrac{1}{n!}$

The TI tells us that this series converges to e-1. Looking at a table, we see in Figure 19.11 that by the 6th term several digits of accuracy are assured. We can arithmetically obtain the value of e to any number of digits of accuracy by adding 1 to the above series and using a sufficiently large n.

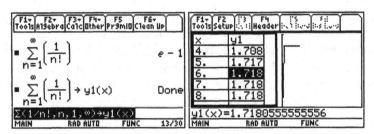

Figure 19.11 Making a graph and table of a fast converging series.

A slow converging series: $1 - \dfrac{1}{3} + \dfrac{1}{5} - \dfrac{1}{7} + \ldots = \displaystyle\sum_{n=1}^{\infty} (-1)^{n-1} \dfrac{1}{2n-1}$

Unlike the last example, some series converge very slowly. You may know a series is convergent, but find getting a highly accurate value can be difficult. In Figure 19.12 we see a graph that looks convergent but the table shows that even after 800 terms we can only comfortably predict that the value is perhaps between 0.78 and 0.79. It turns out that that the true value is $\pi/4 = 0.785398\ldots$ This amazing fact is shown in Chapter 20 by finding the Taylor series for arctan(x). In similar fashion to the previous example we could multiply the series by 4 and have a means of approximating pi to any degree of accuracy. However, in this case, the convergence is so slow that it is not practical.

In general alternating series are not evaluated with exact limits by the TI. This is why we have shown the numerical and graphical approaches in this section. We were lucky to be able to find exact limits in the last two examples, but in general, the limit can be difficult to find.

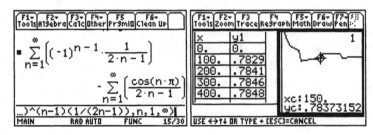

Figure 19.12 A slow converging graph and table.

CHAPTER TWENTY

TAYLOR SERIES

A Taylor series command is built in for producing approximation polynomials. We limit ourselves to confirming a few well-known results and it is assumed that you know the formula for the Taylor polynomial.

Taylor approximating polynomials

The tangent line to $f(x)$ at $x = a$ is considered the best first-degree polynomial that approximates the function near that point. For most functions, a better polynomial approximation is a quadratic model. The best degree n polynomial approximating $f(x)$ for x near a, is called the Taylor polynomial, is given by

$$P_n(x) = f(a) + f'(a)(x-a) + \frac{f''(a)}{2!}(x-a)^2 + \cdots + \frac{f^{(n)}(a)}{n!}(x-a)^n .$$

The TI has a built-in Taylor polynomial generator. In Figure 20.1 we compare our general formula to the built-in function output. The example is the Taylor polynomial of degree 3 for $y = e^x$, centered at 0. We easily verify the Taylor polynomials shown in Figure 20.1 since the function $y = e^x$ is its own derivative and the higher order derivatives are all the same. With a center at $x = 0$, we have $e^0 = 1$ and thus $1/(n!)$ is the coefficient of x^n.

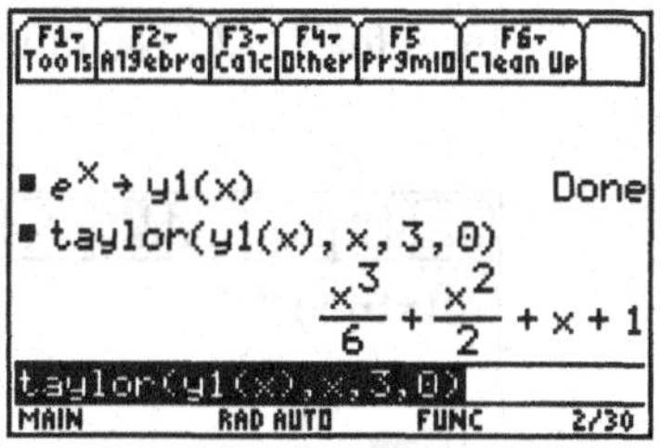

In this chapter we use the built-in taylor command with syntax

taylor (*function, variable, degree, center*).

Figure 20.1 Built-in Taylor polynomial command.

The Taylor polynomials for $y = e^x$

We want to compare the graph of a Taylor polynomial to the original function. For example, we continue $y = e^x$ but find the sixth degree Taylor polynomial with center 1. We can enter the taylor command directly in the Y= editor, shown as y2 in Figure 20.2, but to speed up the graphing we evaluate taylor on the HOME screen and paste the result into y3. We set the graph style of y3 to Path in order to create a little animation.

From the graphs in Figure 20.2 we see that the approximation is quite bad before $x = -1$ and then gives a good fit for the rest of the screen. We can only expect a polynomial to do a good job locally since $y = e^x$ approaches 0 as $x \to -\infty$ and no (nontrivial) polynomial does that. However, this does not mean that for some large negative value of x there is no convergence. Given any value of x, we can find a degree n large enough so the Taylor polynomial centered at 1 provides a close fit. This is not the case for all functions, as we see in the next example where the convergence is limited to a finite interval.

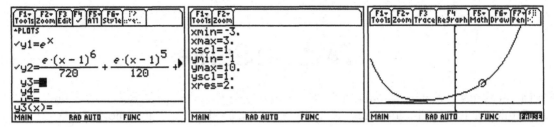

Figure 20.2 A Taylor polynomial of degree 6 centered at 1 approximates $y = e^x$ well for values close to 1.

The Taylor polynomials for ln(x)

As a second example, we find the Taylor polynomials of degree 6, centered at 1 for the function $\ln(x)$. Again, the formula is pasted into y2. The Taylor polynomial graph is shown in Figure 20.3 with a Thick graph style. It can be seen that it is not very good for $x > 2$ and has defined values for $x \leq 0$ where the original function is not defined. However, near $x = 1$ it is an excellent approximation. In the case of e^x, we claimed that for any x-value there was an n for which the Taylor polynomial would converge to the function. For $\ln(x)$ and the Taylor approximations centered at 1, no matter how large n is, the polynomial will never be close to $\ln(x)$ for $x \geq 2$ (The ratio test gives a radius of convergence of 1 about $x = 1$.)

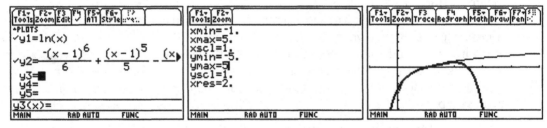

Figure 20.3 The Taylor approximation of order 5 centered at 1 calculated and pasted to y3. The graph window is $-1 \leq x \leq 5$ and $-5 \leq y \leq 5$.

The Taylor approximations of the sine function

In the next example, we increase the degree and see the Taylor polynomials centered at 0 wrap closer and closer to $y = \sin(x)$. In Figure 20.4, only the graphs for degrees 1, 3, 5, 7 are shown since the graphs for degrees 2, 4, 6, 8 (respectively) are the same. This is because the even power Taylor coefficients are zero. Looking at the four graphs, we can see that the approximations are close on wider and wider intervals as the degree gets larger. The interval of convergence for the seventh degree Taylor polynomial seems to be $-\pi < x < \pi$, but increasing the degree shows an increasingly wider convergence. The ratio test can be used to confirm that the interval of convergence for these Taylor polynomials expands to include all real numbers.

> *Tip:* The higher the degree of the polynomial, the longer it takes to graph. You may want to set xres=4.

These approximations of trigonometric and exponential/logarithmic functions are examples of the fact that any infinitely differentiable function can be locally approximated by polynomials. This is important because computers and calculators are masters at polynomial evaluation — after all; it is just addition and multiplication.

Evaluating a Taylor polynomial at $x = 1$ is the same as summing the coefficients. As the degree of the Taylor polynomial goes to infinity, the value at $x = 1$ remains finite because of the $n!$ term in the denominator of the Taylor coefficients. Another way of saying this is that the series of Taylor coefficients converges.

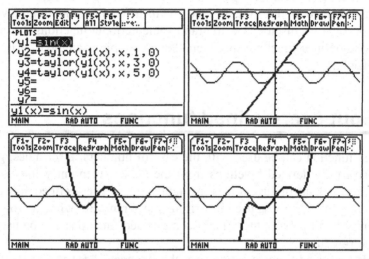

Figure 20.4 The graphs of the Taylor polynomials of degree 1, 3, 5 are shown in bold. The window is ZoomTrig.

The Taylor series of the arctan function

In Chapter 19 we found that the series

$$\sum_{n=1}^{\infty}(-1)^{n-1}\frac{1}{2n-1}=1-\frac{1}{3}+\frac{1}{5}-\frac{1}{7}+\dots$$

converged. In Figure 20.5 we notice that the Taylor polynomials for the function $\tan^{-1}(x)$ have a similar pattern. By setting $x = 1$, we obtain the series in question and knowing $\tan^{-1}(x) = \pi/4$ allows us to conclude:

$$\tan^{-1}(1)=1-\frac{1}{3}+\frac{1}{5}-\frac{1}{7}+\dots=\sum_{n=1}^{\infty}(-1)^{n-1}\frac{1}{2n-1}=\frac{\pi}{4}$$

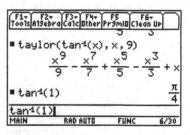

Figure 20.5 The Taylor polynomial of degree 9 shows a pattern used to sum a series.

CHAPTER TWENTY ONE

FOURIER SERIES

The Taylor polynomials were good approximations to a function, but beyond the radius of convergence they were awful. As suggested by the example of the sine function in Chapter 20, Taylor polynomials are practically useless for periodic functions. The global approximation functions called Fourier approximations are used for this but there is insufficient computing power to realistically use the TI for this task. First we discuss how to graph some periodic functions that are not trigonometric, including piecewise defined functions and then show a simple example of a Fourier transform done by the Calculus Toolkit.

A word about user-defined functions

A defined function can be used with the ease of numbers in formulas, graphs and tables. So far we have only defined functions from the HOME screen entry line as a single formula. A more powerful kind of function is available. These user-defined functions can use logical expressions similar to those used in writing a program. But while a program can only be used on the entry line by itself, these more complicated functions can be used within expressions and in the Y= editor.

To define or edit any function, use the Program Editor in the APPS menu. The three options are: 1:Current, 2:Open, and 3:New. We use 3:New in the next example to define a new function within the program editor.

The square wave function

The square wave function is commonly used in electrical engineering to model switching: it is either on or off. We use 1 to mean on, 0 to mean off. The function is defined using the logical structure If…Then…EndIf, which can be pasted as a group from the F2 (Control) menu or separately from the CATALOG. The complete definition is shown in Figure 21.1. The mod function is built into the TI and mod(x,2) returns the remainder after x has been divided by 2.

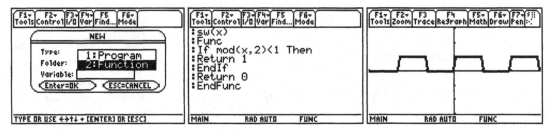

Figure 21.1 A user-defined function definition within Program Editor. The square wave function is graphed as y1=sw(x) in a window setting of -3 ≤ x ≤ 3, -3 ≤ y ≤ 3.

The general formula for the Fourier approximation function

A Fourier approximation function uses the sine and cosine functions to approximate periodic functions. These functions are not polynomials, but we use the polynomial vocabulary to describe them. The term *degree* specifies which sine/cosine terms are included and the specific constant multipliers are called *coefficients*. The definition of the n-th degree Fourier function for the interval $-b/2 \leq x \leq b/2$ is

$$F_n(x) = a_0 + a_1 \cos((\tfrac{2\pi}{b})x) + a_2 \cos(2(\tfrac{2\pi}{b})x) + \ldots + a_n \cos(n(\tfrac{2\pi}{b})x)$$
$$+ b_1 \sin(x(\tfrac{2\pi}{b})) + b_2 \sin(2(\tfrac{2\pi}{b})x) + \ldots + b_n \sin(n(\tfrac{2\pi}{b})x)$$

$$a_0 = \frac{1}{b} \int_{-b/2}^{b/2} f(x)dx$$

$$a_k = \frac{2}{b} \int_{-b/2}^{b/2} f(x) \cos(k(\tfrac{2\pi}{b})x)dx \quad \text{for } k > 0$$

$$b_k = \frac{2}{b} \int_{-b/2}^{b/2} f(x) \sin(k(\tfrac{2\pi}{b})x)dx \quad \text{for } k > 0$$

Since the graphs we will create are for periodic functions, we integrate instead from 0 to b. In cases where the Fourier function is approximating a non-periodic function over an interval of length b, it is important to have the integration centered on the interval.

With the number of integrations and summations involved, it is virtually impossible to use the TI to derive and graph general Fourier functions of any order. We restrict our derivation to a Fourier series of order 1 for the square wave function. A second example is shown from the TI App Calculus Toolkit which has a Fourier option.

A system for graphing the Fourier approximation function

In Figure 21.2 we create and graph the Fourier approximation function of order 1 for the square wave function.

Creating coefficients and defining the function

From Figure 21.1 we see a complete period from the origin has length 2, so three constants needed in the definition of a function of order 1 are:

```
1/2*∫(sw(x),x,0,2)→a0
2/2*∫(sw(x)*cos(2π/2*x),x,0,2)x,→a1
2/2*∫(sw(x)*sin(2π/2*x),x,0,2)x,→b1
```

We enter them, with obvious simplification, on the home screen. Each calculation takes about one minute, which underlines why a computer is needed to create Fourier functions. Next we define the actual function as y2:

```
a0+a1*cos(2π/2*x)+b1*sin(2π/2*x)→y2(x)
```

To test these definitions, we graph the approximating function, y2, with the square wave function, sw(x). Higher orders of Fourier approximations create even better fits.

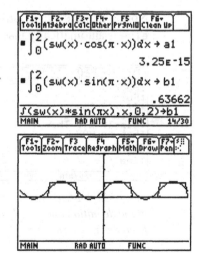

Figure 21.2 A Fourier function of degree 1 fits the square wave function.

> *Tip:*   Numeric approximations of coefficients which are zero often appear as non-zero values with negative exponents of -14 and -15. See the value of a1 in Figure 21.2.

The Calculus Toolkit App

The App, Calculus Toolkit, was previously used in Chapter 11. It has F6 (Advanced), 4:Fourier as an option. A default example is shown in Figure 21.3. Beware that it takes an incredibly long time to produce any examples of functions other than the default.

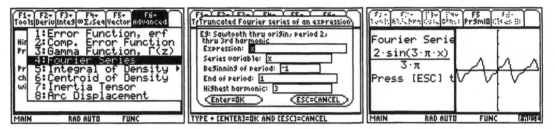

Figure 21.3 Fourier series example using the Calculus Toolkit App.

Piecewise defined functions

The If...Then structure can be extended to more complicated definitions of piecewise defined functions. For example, we next define pwd(x) to have three different formulas used for different values of *x*. Figure 21.4 shows its definition and graph. Notice that a piecewise defined function can be continuous or discontinuous. The last frame shows an alternate means of defining a piecewise defined function. From the Y= editor screen, we use nested logical when(...) commands:

$$y1=when(x<0,cos(x),when(x<1,x^2,1)).$$

Notice that the pretty print feature displays each function on its own line and that the word else is automatically inserted as a separator.

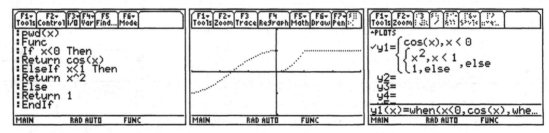

Figure 21.4 If...ElseIf...EndIf structure for a function with three formulas. EndFunc is not shown at the definition end. The last frame shows an alternate means of defining a piecewise defined function from the Y= editor screen.

PART V DIFFERENTIAL EQUATIONS

CHAPTER TWENTY TWO

DIFFERENTIAL EQUATIONS AND SLOPE FIELDS

Situations occur where we know about a rate, but the original function is not explicitly known. Equations that involve derivatives are called differential equations. Most rates are expressed in terms of time, so t is most often the independent variable. The dependent variable is often a quantity, so t and Q are commonly the variables of differential equations. It is also quite common to use the standard x and y variables, with a prime denoting a derivative. For example, two common differential equations are written succinctly as $y' = x$ and $y' = y$. A slope field graph shows local representative behavior of a differential equation and one purpose of this chapter is to introduce this tool.

The solutions of differential equations

We have seen how to antidifferentiate the equation $\dfrac{dQ}{dt} = t$ and solve for Q to find a general solution $Q = t^2 + C$. This means that there is a whole set of solutions which differ by a constant C. However, we often know specific conditions that determine a single solution, called a particular solution. Particular solutions are graphed as curves. It is common to draw a slope field graph to show the general solution and then superimpose a particular solution curve starting at some initial value.

Discrete vs. continuous representation
The TI has a differential equation mode in which a continuous differential equation can be directly entered and graphed. It also has a sequence mode that can enter and graph difference equations — a kind of discrete differential equation. The two approaches, the discrete and continuous, offer insight into the topic when considered together. We start with an example of how a discrete difference equation approximates a differential equation.

The discrete learning curve: a sequence function

One theory of learning is that the more you know, the slower you learn. Let y be the percentage of a task we know. The learning rate is then y' and we assume $y' = 100 - y$. The time unit for this rate is the standard five day work week. The slowing of learning takes place immediately and continuously. We start by looking at a discrete model where we assume the same rate all day. This assumption makes the model discrete.

Start with the data in Table22.1.

Time (working days)	0	1	2	3	4	5	10	20
Percentage learned	0	20	36	48.8	59.0	67.2	89.3	98.8

Table 22.1 Approximate percentage of task learned as a function of time.

Consider a new employee who knows 0% of a task Monday morning at time 0. She learns at the rate $y' = (100 - 0)\%$ during the first day. The part of the task the employee learns on Monday, one fifth of a work week, is

$$y' \cdot (1/5) = 100\%(0.2) = 20\%$$

At this rate, she would have the task entirely mastered in a week. But Tuesday, because she already knows 20% of the task, her learning slows to

$$y' = 100\% - 20\% = 80\%$$

Therefore, on Tuesday the part of the task she learns is an additional

$$y' \cdot (1/5) = 80\%(0.2) = 16\%$$

In total, she has learned 20% + 16% = 36% by the end of Tuesday.

Each daily total depends on the previous day's total. Writing this as a discrete function, that can be entered in the TI, the n-th total is

$$u1(n)=u1(n-1)+(1/5)(100-u1(n-1))$$

Making a sequence table

To create a table of this data, we enter MODE and change the Graph setting to 4:SEQUENCE. Using the Y= editor shows functions named u1, u2, ...u99, and below each function is its initial value variable ui1, ui2, ...ui99. In Figure 22.1 the function u1(n), listed above, and its initial value of 0 have been entered.

A table (♦_TABLE) is shown starting at 1 with a step of 1. The highlighted table value means that the employee learns only ≈67% of the task by the end of Friday. Scrolling down shows, according to the model, it is impossible to completely learn the task. If your table does not match the one in Figure 22.1, you may need to set the WINDOW with values shown in Figure 22.2.

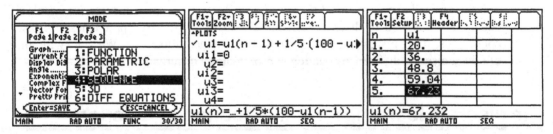

Figure 22.1 The Graph option SEQUENCE in the MODE menu changes the function definitions to u1.

Tip: With the graphing mode as SEQ (shown on the status line), the TblSet values define with tblStart and Δtbl are for n instead of x.

Making a sequence graph

The graph for a sequence function is produced in the same way as for a "normal" y function. We already have u1, so all we lack is setting the window. In Figure 22.2 you see an added set of window variables needing to be set. In this case we make the *n*-values match the *x*-values. The sequence mode graph is a set of discrete points. The values are traced with F3 (Trace).

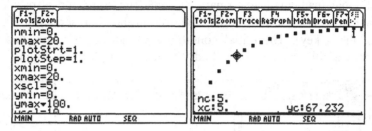

Figure 22.2 Graph (with trace) of the learning sequence function for 20 days.

> *Tip:*   Each graphing mode has its own window settings. Changing xmin in SEQUENCE mode does not change xmin in FUNCTION mode.

The continuous learning curve $y' = 100 - y$

The discrete version of the learning curve assumed that the worker had the same learning rate all day. Now suppose that the learning rate changed every instant, a continuous function. In Figure 22.3, we use MODE to change the Graph type to 6:DIFF EQUATIONS. As with other graph types, we enter the equations with the Y= editor. There we see the differential equations are defined using y1', y2', ... and corresponding initial values yi1, yi2, ... In setting the window the *t* variable for differential equations is like *n* for the sequence equations. For our graph window settings, we make the *t*-values the same as the *x*-values. Recall that this differential equation has weeks as its unit so that xmax=4 gives the same graph as the sequence setting with xmax=20 (since 4 weeks = 20 workdays).

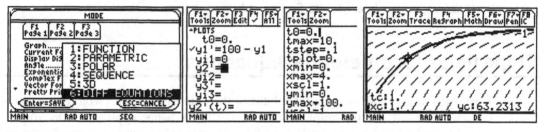

Figure 22.3 The continuous learning curve and a particular solution with y(0)=0.

> *Tip:*   In defining y1'=, use y1, not just y.

> *Tip:*   Your graph in Figure 22.3 may not show the set of small lines superimposed on the graph, these are slope lines. This is explained shortly.

The analytic solution of a differential equation

You may ask, what is the equation of the graph shown in Figure 22.3? That is the essence of solving a differential equation. The TI finds this kind of symbolic solution in most cases. Unlike the yi' variables in the Y= editor, we enter differential equations on the HOME screen in the traditional mathematical format. The deSolve command is pasted from the F3 (Calc) menu. When entering a differential equation on the HOME screen, use the ' symbol (the 2nd mode of the = key). The initial conditions are added with and, and the independent and dependent variables must be identified. This format is shown in Figure 22.4.

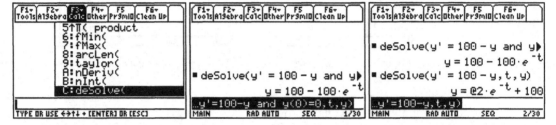

Figure 22.4 The deSolve in the F3 men is used to find the particular solution of the learning curve's differential equation and its general solution. The entry line shows the syntax that follows after the deSolve beginning.

What if we forget or do not know the initial conditions? This leads to the situation mentioned at the beginning of the chapter where a general solution $y = t^2 + C$ that includes a constant. The TI identifies such constants by @1, @2, etc., similar to what we have seen in some solutions involving periodic functions. The symbol @ is the ♦ version of STO; it is also in the 2nd_CHAR, Punctuation menu if you need to use it.

Tip: You need not be in the DIFF EQUATION graph mode to use deSolve.

Summary of three approaches to differential equations

We have seen three approaches to differential equations. First we used difference equations (SEQUENCE graph mode) to make table values to approximate solutions. Second, we used DIFF EQUATION graph mode, to enter the equation and graph a particular solution. Finally, from the HOME screen, we used deSolve to find a symbolic solution.

Slope fields for differential equations

If we tried to graph all the general solutions, then the entire screen would be covered as our constant changed and the graph shifted vertically. A better approach is to graph selected particular solutions. A common practice is to refine this further and graph discrete linear pieces that approximate a particular solution. We call such a graph the slope field of the equation. The setting to display the slope field of a differential equation graph is on the GRAPH FORMATS menu, accessed by ♦_| (Or F1 Tools, 9:Format). Figure 22.5 shows selecting the SLPFLD option and the graph of our previous example.

Tip: The Fields option only appears in the GRAPH FORMATS menu when the graph mode is DIFF EQUATION.

From the slope field, we trace out particular solutions. For example, it is easy to find the learning curve for an employee with 50% of the skills hired at the beginning of the second week. Pressing F8 (IC) activates a circular cursor that can be moved to the point on the screen where you want a particular solution to begin.

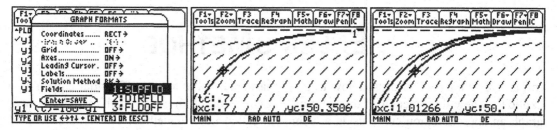

Figure 22.5 Slope field for the learning curve. Use F8 (IC) to draw additional particular solutions through the point (1,50).

Slope fields examples for differential equations

As a beginner to differential equation graphing it is suggested that you enter an equation only in y1'=. If the equation is in traditional form, translate x to t and y to y1, as shown below.

Slope field for $y' = y$

In the Y= editor, enter y1'=y1. Use a ZoomDec window for the next three examples. The solution graphs are exponential of the form $y = e^x + C$. Your screen may appear with fewer slope lines, the last option in the WINDOW editor was set as fldres=18.

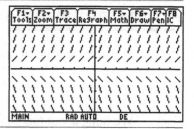

Slope field for $y' = 2x$

In the Y= editor, enter y1'=2t. Use the previous window. The solution graphs are exponential of the form $y = x^2 + C$.

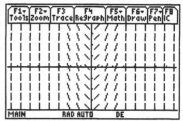

Slope field for $y' = -x / y$

In the Y= editor, enter y1'=-t/y1. Use the previous window. The solutions are circles centered at the origin. By implicit differentiation, we know that this differential equation is obtained from $x^2 + y^2 = a$ where $a > 0$.

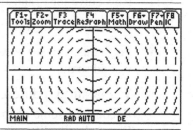

Slope field $y' = (-y + x*y) / (x - x*y)$

In the Y= editor, enter y1'=(-y1+t*y1)/(t-t*y1) and a window: $0 \leq x \leq 3$ and $0 \leq y \leq 3$. A particular solution is an oval curve, unless we are at (1,1), a fixed point. This predator-prey model equation is discussed in Chapter 25.

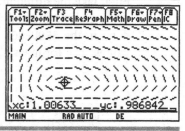

CHAPTER TWENTY THREE

EULER'S METHOD

In the last chapter we discussed how to graph particular solutions to differential equations when we explicitly knew the function of the solution. When we don't know the explicit formula, we can still graph an approximation curve using a technique called Euler's method. The general idea of Euler's method is this: "From a specific starting place, follow the slope line through this point for a short time. Then stop and find a new slope line at the current point and follow that line for a short time. Continue in this manner." We relate this to a differential equation and use the sequence mode to do our "stop after a short time" calculations. If better accuracy is needed, then we stop more often. We apply this technique to particular solutions of several differential equations.

A differential equation in relation to a difference equation

Suppose we want to find the solution of the differential equation

$$\frac{dy}{dx} = F(x, y) \text{ starting at the point } (x_0, y_0)$$

We think about this continuous differential equation in a discrete way and write

$$\frac{\Delta y}{\Delta x} = F(x, y)$$

or, more explicitly,

$$\frac{y_n - y_{n-1}}{\Delta x} = F(x_{n-1}, y_{n-1}) \text{ with } \Delta x = x_n - x_{n-1}$$

Now solve for y_n and x_n:

$$y_n = y_{n-1} + F(x_{n-1}, y_{n-1})\Delta x, \qquad x_n = x_{n-1} + \Delta x$$

From this, you see how knowing (x_0, y_0) with $n = 1$ allows us to find y_1. We then use y_1 to find y_2, and so on.

Translating a differential equation to a difference equation

Reconsider some of the differential equations from the last chapter as difference equations. First, take the differential equation $dy/dx = y$ starting at $(0, 1)$. With $\Delta x = 0.1$, we make a table using Euler's method and then make a graph to compare it to our analytic solution. This is done in the SEQUENCE graph mode. We use u1 to store the x-values and u2 to store the y-values. The formulas above translate to

$$x_n = x_{n-1} + \Delta x \text{ becomes } \text{u1(n)=u1(n-1)+.1}$$

$$y_n = y_{n-1} + F(x_{n-1}, y_{n-1})\Delta x \text{ becomes } \text{u2(n)=u2(n-1)+u2(n-1)(.1)=1.1*u2(n-1)}$$

Figure 23.1 shows the Y=, WINDOW and TABLE setup for finding the u2 values with $\Delta x = 0.1$.

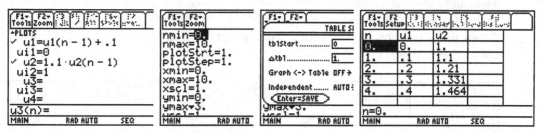

Figure 23.1 Making a table for the approximation by Euler's method to the particular solution of the differential equation y' = y.

We set u3=e^(n/10) to be the exact solution we found in the last chapter, $y = e^x$. (Note that n/10 is used since $\Delta x = 0.1$.) We compare the numeric approximations in u2 to the true values in u3 by using TABLE. The Euler method approximation for an *x*-value such as 0.3 is pretty good, but you can see the error increase as *x* increases. If needed, these approximations can be improved by making Δx smaller. In Figure 23.2 we graph u2 as square data points and u3 as a line for comparison. We see the approximations starting to be low at $n = 4$.

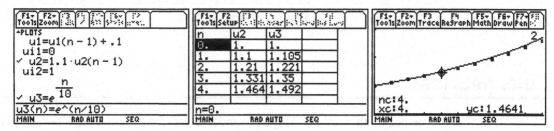

Figure 23.2 Making a sequence table and graph to compare the approximation by Euler's method to the particular solution of the differential equation y' = y. Use Trace to identify values.

Euler's method for *y'* = *-x* / *y* starting at (0,1)

The TI knows two methods for solving differential equations numerically. The default setting is RK which stands for the Runge-Kutta method. This is slower than the Euler method, but it is generally more accurate (see any differential equations text for details). The alternate setting is EULER.

Consider the differential equation $dy/dx = -x/y$ starting at (0,1). We investigate the effect of using a smaller value of Δx by graphing the solution on the interval $0 \le x \le 1$ first with $\Delta x = 0.1$, then with $\Delta x = 0.01$. We use the built-in Euler setting. The setup process for the screens in Figure 23.3 follows these steps:

    1. In MODE, change the Graph setting to 6:DIFF EQUATIONS

    2. Press ♦_Y= and F1 8:Clear Functions.

    3. Enter the equation: y1'=-t/y1 and initial condition yi1=1.

    4. Press ♦_| (or F1 Tools, 9:Format) to access the GRAPH FORMATS and set the Solution Method to 2:Euler and set Fields to 3:FLDOFF.

    5. Set the window as shown in Figure 23.3

We are interested in $0 \leq t \leq 1$. With the variable change from x to t, now Δx = tstep=.1. To insure a circular graph, the aspect ratio of the window is set to 2:1. The iterations between each tstep is given by Estep; 1., is the default. After the first graph and trace, increase the Estep to 10 to improve the accuracy of the approximation. Notice that the graph is more circular, as it should be.

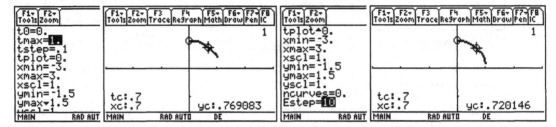

Figure 23.3 Euler's method to the particular solution of the differential equation y1'=-t/y1 and initial conditions yi1=1. First with Estep=1 and then Estep=10. Use Trace to identify values.

Tip: Using tstep=.01 with Estep=1 produces the same numeric values as tstep=.1 with Estep=10., so we did in effect graph the solution with $\Delta x = 0.01$. Changing Estep instead of tstep makes graphing and tracing faster.

Euler gets lost

Use Euler's method with caution. As you wander away from your initial point, you may encounter increasing errors. The next example shows a path where Euler encounters an infinite slope that causes problems.

Let's take the previous example and increase tmax to 2 in hopes of completing the circle in the first quadrant. Figure 23.4 shows the folly as we come to (1,0). Changing the setting from Euler back to the default RK eliminates this hazard as shown in the last frame of Figure 23.4. Unless investigating Euler's method, use the RK setting.

Figure 23.4 With tmax=2, Euler's method breaks down at (1,0) where the slope is undefined. This problem disappears after changing Solution Method to RK.

CHAPTER TWENTY FOUR

THE LOGISTIC POPULATION MODEL

In this chapter we derive a logistic equation to model the population growth of the United States. The exposition here is a brief introduction to the TI statistical capabilities, which are a powerful application of this calculator.

Using the Data Editor to enter US population data

The first step is to enter the data given in Table 24.1, from the US Census Bureau, into two lists. Typically, annual data is not indexed by the year number itself, but by years since a base year. Our base year is 1790 and the populations are rounded to the nearest tenth of a million. You may wonder why we don't include the most current census data, the answer is that we want to use these earlier years to make a population prediction equation and then compare its prediction with the actual current data.

Year	1790	1800	1810	1820	1830	1840	1850
Years since 1790	0	10	20	30	40	50	60
Population ($\times 10^6$)	3.9	5.3	7.2	9.6	12.9	17.1	23.1

1860	1870	1880	1890	1900	1910	1920	1930	1940
70	80	90	100	110	120	130	140	150
31.4	38.6	50.2	62.9	76.0	92.0	105.7	122.8	131.7

Table 24.1 US Census Data 1790 – 1940.

To create a new data file, press APPS, 6:Data/Matrix Editor, 3:New…. If you are using the Apps Desktop, then Data/Matrix Editor is a desktop icon. The Data Editor presents a dialog box, shown in the middle frame of Figure 24.1, which is like the one used for programs and functions.

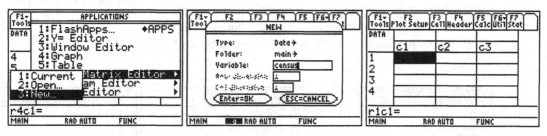

Figure 24.1 Creating a new data table.

Dialog boxes require up and down arrows to move among items and the right arrow to produce item choices. In the dialog box we type a variable name. You must press ENTER after making an entry, then you must press ENTER again before you exit to OK all the current dialog box settings. After naming the data set census, you are presented an empty table with columns c1, c2, c3.

The years can be entered sequentially, but it is easier to press F4 (Header) and enter the seq command shown on the entry line of the first frame in Figure 24.2. Arrow to the top Title row to enter a text name to identify the column. Enter the population data in column c2.

F1▾ Tools	F2 Plot Setup	F3 Cell	F4 Header	F5 Calc	F6▾ Util	F7 Stat
DATA						
	c1	c2	c3			
1	0					
2	10					
3	20					
4	30					
c1=seq(10*x,x,0,15)						
MAIN	RAD AUTO	FUNC				

F1▾ Tools	F2 Plot Setup	F3 Cell	F4 Header	F5 Calc	F6▾ Util	F7 Stat
DATA	year					
	c1	c2	c3			
1	0					
2	10					
3	20					
4	30					
c2,Title=pop						
MAIN	RAD AUTO	FUNC				

F1▾ Tools	F2 Plot Setup	F3 Cell	F4 Header	F5 Calc	F6▾ Util	F7 Stat
DATA	year	pop				
	c1	c2	c3			
1	0	3.9				
2	10	5.3				
3	20	7.2				
4	30	9.6				
r4c2=9.6						
MAIN	RAD AUTO	FUNC				

Figure 24.2 Entering data using seq and labeling columns in a data table.

Tip: When returning to the last used data file, you can press APPS 6 ENTER. If you are looking for a specific data file, use APPS 6 2:Open to find it.

Using Plots to graph data lists

To look at the data graphically, we press F2 (Plot Setup) and to reach the dialog box shown at the top of Figure 24.3. This dialog box holds a summary of plot assignments for graphing data lists. Here plots are selected/deselected and, for each plot, it is the entryway to the next screen in which actual definitions are made. Press F1 (Define). Arrow down to x and type c1 ENTER, then arrow to y and type c2 ENTER. Press ENTER again to save these settings.

In Figure 24.4 we look at the Y= editor to be sure that other functions are turned off and you can see that Plot 1 is turned on. To set a graph window press F2 (Zoom). A handy zoom setting is 9:ZoomData which sets a window that fits all your data. The ZoomData selection automatically graphs the data points. You can also use the usual menus at the top to trace and investigate the graph. If there are other graphs that appear, you may not have turned off all the Y= functions. Use F5 (All) 3:Functions Off, if needed.

Figure 24.3 Plot Setup, F2 (Plot Setup) and F1 (Define).

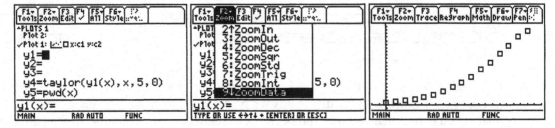

Figure 24.4 Scatter Plot using ZoomData.

Regression equations

Another statistical feature of the TI calculator is its ability to find regression equation models of various kinds. The word regression is a statistics term, but for our purposes think of it as meaning a *best fit* type of equation. The linear regression equation is the most common, it gives a line that comes closest to graphically fitting our data. We also have exponential regression, giving an exponential equation that best fits the data. But, our scatter plot in Figure 24.4 has a logistic curve look, so we fit it using a logistic regression equation.

The long road to a logistic graph

The number of screens traveled through in order to calculate and graph the logistic regression looks huge, but in fact the work flow goes quickly. All nine screens are in Figure 24.5 and correspond to these nine steps 1-9.

1. Press APPS 6 1 to return to the original data screen.
2. Press F5 (Calc) and arrow right to see the Calculation Type menu.
3. The regression choices are from 3 to C. Choose C:Logistic.
4. Enter c1 as x and c2 as y. Arrow down to Store RegEQ to.
5. Arrow right and select y1(x) to store the derived equation after it is calculated.
6. Setup is complete as shown. Press enter and wait for calculation.
7. Note the coefficients. Press ENTER.
8. Press Y= to check that the equation was pasted and that it and Plot 1 are selected.
9. Press ♦_GRAPH to see the plot and graph. It uses the previous ZoomFit window.

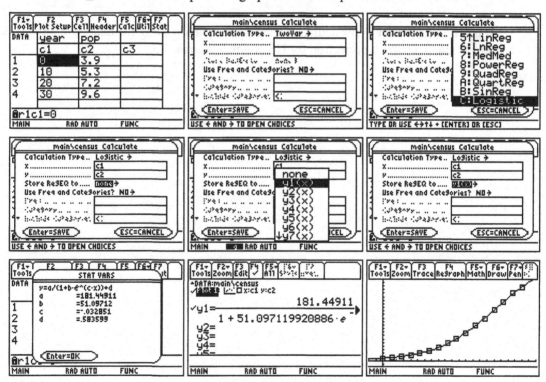

Figure 24.5 Nine steps to finding a regression equation and plotting the graph.

Tip: The various STAT CALC regression options allow you to try out different models on the same data to look for a good equation.

Our logistic model equation is

$$P = \frac{181}{1 + 51e^{-0.03285t}}.$$

It turns out that this equation is not a very good model for the current population because of the unexpected baby boom. The data for years 1950 to 2000 is added and shown in Figure 24.6, along with the regression equation graph from 1790 to 1940. The difference between the predicted and actual is dramatic.

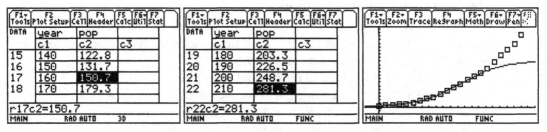

Figure 24.6 Adding data for 1950 (150.7) to 2000 (281.3) and showing the old regression equation.

Tip: If a formula was used to generate c1 in the census data table, then the cells in that column are locked against editing. Amended the upper limit of the header formula to be seq(10*x,x,1,21).

It is left to the reader to recalculate a regression equation using the data through the year 2000 and to graph a current projection for US population growth.

CHAPTER TWENTY FIVE

SYSTEMS OF EQUATIONS AND THE PHASE PLANE

In this chapter we look at examples of systems of differential equations where the independent variable is time. These equations are graphed to show their relationship to each other. Points on the graph are called the phase trajectory or orbit in the phase plane. Two popular examples of using systems of differential equations are the S-I-R model the predator-prey model.

The S-I-R model

The S-I-R stands for Susceptible - Infected - Recovered, so you can tell that this is used to model epidemics. The population is divided into the three groups and people move from S to I to R, or they just stay in S. The three rates in terms of time are

$$\frac{dR}{dt} = bI$$ the recovery rate depends upon the number infected

$$\frac{dS}{dt} = -aSI$$ the susceptible rate is negative and depends on both the number of infected and susceptible

$$\frac{dI}{dt} = aSI - bI$$ the infected rate is the negative of the sum of the other two rates.

Knowing any two of the quantities S, I, R, automatically determine the third, so we concentrate on the last two rates.

The boarding school epidemic

The following simple example is used to illustrate the S-I-R model. There were 762 students in a boarding school and one returned from vacation with the flu. Two more students became sick the second day. From this we know

$$S = 762 \text{ and } I = 1 \text{ and } \frac{dS}{dt} = -2$$

and we find $a = -0.0026$ from

$$\frac{dS}{dt} = -aSI \text{, where } (-2) = a(762)(1)$$

This flu lasts for a day or two, so we assume half of the sick get well each day, thus, $b = -0.5$.

Time plots for the model

In Chapter 23 we listed steps to set up a differential equation graph. In this chapter we sometimes need another step to change the assignment of the axes. After defining the equations, use F7 (Axes) to check, and if necessary, change the x- and y-axes. In all previous examples we used the default setting TIME (the axes are t and y), but after the first example, in which we use TIME, we will change the setting from TIME to CUSTOM. Because the axes can be custom set in the DE mode, we identify them by setting Labels to ON with ♦_|. The graph steps are now:

1. Set MODE to DIFF EQUATIONS
2. Define the equation with Y= editor (include initial conditions) Set other functions off
3. **New:** Set AXES with F7 (Axes) to either TIME or CUSTOM
4. From the Y= editor, set GRAPH FORMAT options with ♦_|
5. Set WINDOW
6. Draw GRAPH

The six screens in Figure 25.1 follow these six steps. We define the differential equations and graph the variables S and I as y1 and y2 over time. With the default axes set to TIME, both y1 and y2 are graphed against time on the horizontal axis. With two or more functions, slope fields cannot be drawn, so we use ♦_| to select Fields: FLDOFF. We also turn the axes Labels ON. For the window, we look at the epidemic for 20 days on the horizontal axis and set ymax=800, as there are just under 800 students. Tracing gives an estimate for the maximum number of infected students.

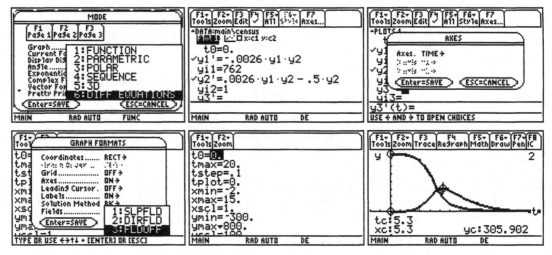

Figure 25.1 The graphs of S and I over time. We see that the susceptible are in a constant decline and the infected peak in the sixth day.

Phase plots for the model

In Figure 25.2 we look at the *SI* phase plane. The same definitions of y1' and y2' and initial conditions remain from the previous plot, but now we change the AXES to CUSTOM and assign: X Axis:y1 and Y Axis:y2. The model requires $0 \le y1 \le 800$, $0 \le y2 \le 400$ and $0 \le t \le 20$ days. Watching the screen as it graphs, you see the graph drawn from right to left, since the first point is $(S, I) = (762, 1)$. Using TRACE on this phase plot feels backward because the right arrow key increases t, causing the trace cursor to move left. If you press the left arrow key at the start, then nothing happens. TRACE shows the peak at approximately $(192, 306)$, where $t = 5.3$ days. This maximum value is of interest in the next section.

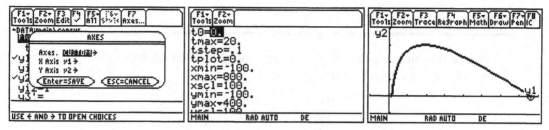

Figure 25.2 The SI phase plane. Use F7 (Axes) in the Y= editor for CUSTOM settings.

> **Tip:** The TRACE cursor moves through the points connected to the *t*-values when the right arrow is pressed; the cursor movement is not necessarily from left to right. Moving backward through time is much slower than moving forward.

> **Tip:** Pressing the 2nd key before the right arrow puts it in turbo mode, making the cursor jump further.

Direction fields for the S-I-R model

Like a slope field for a function in terms of *t*, we graph slope lines for the relationship of *I* to *S*. The direction field shows a relationship between two quantities, like y1 and y2, in a system. Time is not shown on either axis, but each point does have a time value implicitly attached. The time value is seen when you trace a phase plot; the cursor starts at tmin and increases by tstep.

We use ♦_| to set Fields: DIRFLD. The resulting graph is shown in Figure 25.3. Of particular interest is the fact that along any particular solution, the peak (threshold value) is at the same value (*S* = 193) on the horizontal *S*-axis. But if *S* is less than 193, then the *I*-values decrease immediately. We use F8 (IC) to explore another phase plot at a smaller school (y1 = 500) with one student (y2 = 1) beginning the outbreak.

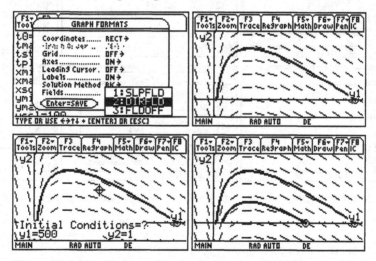

Figure 25.3 The SI phase plane with direction field. Use F8 (IC) to graphically explore particular solutions with other initial conditions.

> **Tip:** Pressing F8 (IC) activates a circular cursor to move to the initial conditions, but it is much easier and more accurate to type in the two desired values.

Predator-prey model

Let r be the number of robins (predators) and w the number of worms (prey). We start with the well-known *Lotka-Volterra* equations:

$$\frac{dw}{dt} = aw - cwr \text{ and } \frac{dr}{dt} = -br + kwr$$

For our simple purposes, we simplify by setting the parameters a, b, c, and k to all be 1.

The simplified predator-prey system is

$$\frac{dw}{dt} = w - wr \text{ and } \frac{dr}{dt} = -r + wr$$

For the TI we define the predator population r as y2 and the prey population w as y1 (with units in millions). This translates to

$$\text{y1'=-y1+y1*y2 and y2'=y2-y2*y1}$$

We follow the presentation order of the previous model and first graph the two populations over time.

Time plots for the predator-prey model

In Figure 25.4 the individual predator and prey differential equations with respect to time are entered. Initially, you may have no idea how to set the window, but a little trial-and-error leads to the window shown in Figure 25.4. It is wide enough to show the periodic nature of the two graphs.

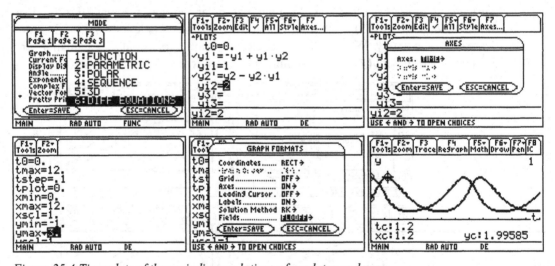

Figure 25.4 Time plots of the periodic populations of predator and prey.

Tip: The STYLE for DE graphing is THICK. You can set it to Line for one of the two equations to distinguish them without having to use Trace.

Phase plots for the predator-prey model

In Figure 25.5, we change the AXES and set the window to match the size of the predator and prey numbers. From the previous graph, we know that more than a complete cycle takes place if tmax=12, so we leave that setting. Or, we could estimate a smaller value of eight by looking from peak to peak on the graph. Set xmax=3 since both populations are less than 3 (million). Remember when tracing values to press the right arrow key even though the trace cursor moves left at first.

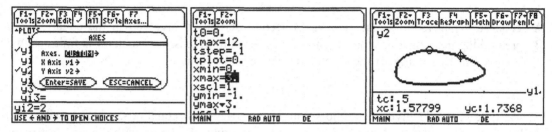

Figure 25.5 Graphing prey against predator.

Tip: When graphing a phase plot, the curve has the style specified by the function defined on the *x*-axis.

Direction field for the predator-prey model

Use ♦_|, to change the Fields setting to DIRFLD, we superimpose the direction field over the phase plot. What we see from this direction field in Figure 25.6 is an equilibrium point at (1, 1). At these values, the populations theoretically remain stable and do not have the cycles we saw in Figures 25.4 and 25.5.

The direction field of this system of two differential equations might look familiar; it was part of the last example in Chapter 22. There it was identified as the slope field for

$$\frac{dy}{dx} = \frac{-y+xy}{x-xy}.$$

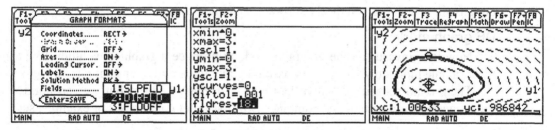

Figure 25.6 The direction field over the phase plot of Figure 25.5.

CHAPTER TWENTY SIX

SECOND-ORDER DIFFERENTIAL EQUATIONS

A second-order differential equation has a second derivative involved in the equation expression. Before making the transition from a symbolic equation to necessary calculator definitions, we rewrite these equations by solving for the second derivative in the form

$$y'' = F(x, y, y').$$

As before, the TI gives symbolic solutions of differential equations, when possible, but it also finds numerical solution values and a graph of the solution. We begin with the simplest second order equation.

The second-order equation $s'' = -g$

A classic second-order equation from physics is,

$$\frac{d^2s}{dt^2} = -g$$

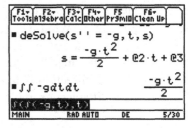

where g is the constant force of gravity on a falling object, s is displacement in feet, and t is time in seconds. If we assume the initial velocity is zero ($v_0 = 0$) and the initial distance is zero ($s_0 = 0$), antidifferentiating once gives $s' = -g \cdot t$, and then again gives the solution $s = -g \cdot t^2/2$. The first calculation in Figure 26.1 uses deSolve, which gives the general answer with arbitrary constants like @2 and @3. In our case both are zero. The second calculation confirms our answer and shows that the TI allows nested integration.

Figure 26.1 The TI allows nested integration and solves second-order equations.

DE graph mode

Now let's investigate the DE graph mode to produce a graphical solution. We also make this example more down to earth by assuming that the gravity constant g is 32 feet/sec^2.

There is no second derivative like, y1''= in the Y= editor, but we can build a second-order equation as a system that combines two first-order equations. For example, set

$$y1'=y2 \text{ and } y2'=-32.$$

So,

$$y1'' = (y1')' = (y2)' = y2' = -32$$

Tip: The second-order equation entry for DE graphing is much more complicated than the straightforward entry of an equation in the deSolve command.

After defining the second-order equation as a set of two first-order equations, deselect all but the first function to be graphed. In Figure 26.2 the equations are entered and y2' is turned off since it is an intermediate function; the solution being graphed is y1, so y1' is selected. Use ♦_| to set the graph formats as shown. Next, set a good window to look at the graph for $0 \leq x \leq 2$. Since the solution is $y = -16x^2$, set the window to show mostly negative y-values. Once the graph is drawn, you can use DRAW, TRACE or F8 (IC) to investigate further.

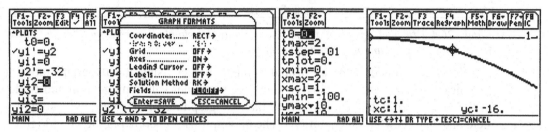

Figure 26.2 Graphing the solution of a second-order equation entered as a set of first-order equations.

> **Tip:**   From the window screen it is possible to know the Solution Method. If diftol is shown as the last setting, the method is RK (the value prescribes the error tolerance in the Runge-Kutta algorithm), otherwise EStep (Euler step) is shown.

The second-order equation $s'' + \omega^2 s = 0$

Next we look at a second-order differential equation that depends only on the dependent variable.

$$\frac{d^2 s}{dt^2} = -\omega^2 s, \text{ where } \omega > 0$$

This differential equation describes simple harmonic motion and the ω (omega) is a constant that is commonly shown in this general equation. To show the capability of the TI to write Greek and International characters, we enter the above equation using ω. We see in Figure 26.3 that its general solution is

$$s(t) = C_1 \cos\omega\, t + C_2 \sin\omega\, t$$

As a specific example, let $\omega = 2$ and set the initial conditions to be $s(0) = 1$ and $s'(0) = -6$. In deSolve, the initial conditions are place inside the deSolve command and are joined to the general equation by and. (Use and from the CATALOG.) The complete entry is

deSolve(s''=-4s and s(Ø)=1 and s'(Ø)=-6,t,s)

We see in the final frame that the particular solution has constants $C_1 = 1$ and $C_2 = -3$.

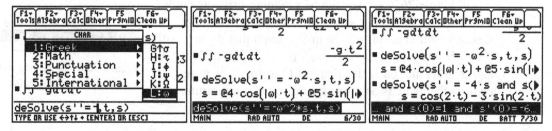

Figure 26.3 A general and a specific solution of a second order equation.

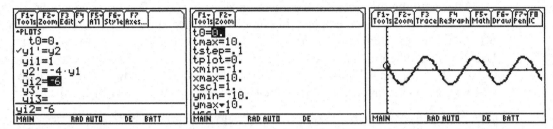

Figure 26.4 A DE mode graph of a particular solution for simple harmonic motion. The ZoomStd window for DE graphing is defined to be off-center.

You may wonder why we would want to bother using the DE graph mode to find a graphical solution when we already know the particular solution equation and could graph it in the FUNCTION mode. The answer is that, by graphing in DE mode, we can add the slope field and enter new initial conditions. The DE graph with ZoomStd is shown in Figure 26.4.

The linear second-order equation $y'' + by' + cy = 0$

Equations of this type are called *linear* second-order equations because isolating the second derivative gives a linear equation with variables y and y' and coefficients b and c. One application of this general equation is describing the motion of a spring.

In Figure 26.5 we solve for the general solution of this second-order linear equation. The solution hinges on the sign of the $b^2 - 4c$ inside the square root in the exponent. The solution shown in Figure 26.5 can be written as

$$y = C_1 e^{r_1 t} + C_2 e^{r_2 t} \text{ with}$$

$$r_1 = \frac{\sqrt{b^2 - 4c}}{2} - \frac{b}{2} \text{ and } r_2 = \frac{-\sqrt{b^2 - 4c}}{2} - \frac{b}{2}$$

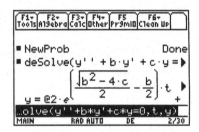

Figure 26.5 The general solution of the linear second-order equation.

These two values r_1 and r_2 are solutions of the equation, $r^2 + br + c = 0$. This quadratic is called the *characteristic equation* for the differential equation. We consider the three possible cases: $b^2 - 4c > 0$, two real distinct solutions; $b^2 - 4c = 0$, one real (repeated) solution; and $b^2 - 4c < 0$, two solutions with imaginary parts.

The overdamped case: $y'' + by' + cy = 0$ with $b^2 - 4c > 0$

Consider the second-order linear equation with $b = 3$ and $c = 2$, so that $b^2 - 4c > 0$ is positive. The general solution and a particular solution with initial conditions, $y(0) = -.5$ and $y'(0) = 3$, is shown in Figure 26.6.

In Figure 26.7, we see the DE graphing using as before a system of first-order equations. Thinking of this as a spring's motion in oil, we see in the graph that it starts half a unit below equilibrium, swings past equilibrium ($y = 0$) and then its motion is damped until the spring is at rest. It appears that there is only one swing across equilibrium.

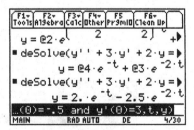

Figure 26.6 The characteristic equation has distinct real roots.

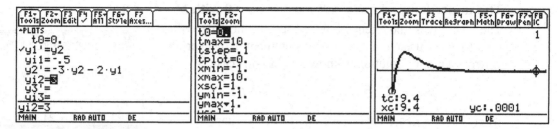

Figure 26.7 A DE graph of a particular solution.

How can we be sure that the spring does not oscillate through equilibrium as it comes to rest? From a DE graph, the Zero option is not available, but using F2 (Algebra), 4:zeros you can confirm that there is a single zero at *t* = 0.223144.

The power of DE graphing is that we can explore new initial conditions. In Figure 26.8, we press F8 (IC) and are met with a new dialog box; just press ENTER and a circular cursor allows you to set new initial conditions. By moving the cursor to (0,0) and pressing ENTER, a second graph is drawn. In this case, the spring starts at equilibrium and appears to never swing back past the equilibrium as it comes to rest.

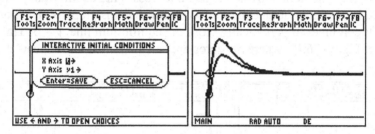

Figure 26.8 Use F8 (IC) to set a new initial condition, y(0)=0.

The critically damped case: *y″* + *by′* + *cy* = 0 with *b²* – 4*c* = 0

To find the general solution when $b^2 - 4c = 0$, we return to deSolve and our original equation with the added restriction that $c = (b/2)^2$. The entry line is

$$\text{deSolve(y''+b*y'+c*y=0,t,y)|c=(b/2)^2}$$

This gives the general solution shown in the first frame of Figure 26.9. For a specific example, suppose *b* = 2 and *c* = 1, then deSolve gives a solution that we can enter in the Y= editor and obtain a graph of the particular solution. With the same initial conditions, *y* = -.5 and *y′* = 3 at *t* = 0, and window as in the last example, the graph is similar to the solution of the previous case, but notice that it is not damped as quickly. Use trace to confirm that it passes through equilibrium three times in the window shown.

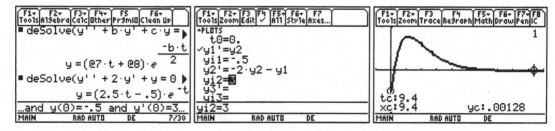

Figure 26.9 The critically dampened graph, similar to the overdamped case of Figure 26.7.

Tip: Don't forget to deselect y2' after editing it when reproducing Figure 26.9.

The underdamped case: $y'' + by' + cy = 0$ with $b^2 - 4c < 0$

The TI gives no general solution if the inequality $c > (b^2)/4$ is used as a conditional statement. However, the general symbolic solution is known and in this case is

$$y(t) = C_1 e^{\alpha t} \cos \beta t + C_2 e^{\alpha t} \sin \beta t,$$

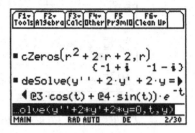

where $r = \alpha \pm i\beta$ are the complex zeros of the characteristic equation.

Figure 26.10 Verify the general solution for a particular case when a=b=2.

When the parameters b and c are given real values, deSolve does produce the solutions. For example, suppose $b = 2$ and $c = 2$, then the $b^2 - 4b < 0$. Figure 26.10 shows an example of such a solution. First use cZeros to find $\alpha = -1$ and $\beta = 1$. Next, use deSolve to verify the symbolic solution displayed above. It is the same, with $C_1 = @3$ and $C_2 = @4$ and the e^{-t} factored to the right.

For comparison to the other two damped cases, in Figure 26.11 we use DE graph mode to show a particular solution with the previous boundary conditions in the same window. Using trace, we see that the motion is damped but crosses the equilibrium point several times in that interval. From the solution equation, we know that this graph is an exponentially damped sine curve. It continues to periodically cross the equilibrium line, $y = 0$. This is verified in the last frame of Figure 26.11, where the zeros are calculates and @n1 appears in the solution.

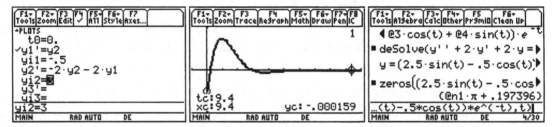

Figure 26.11 Graph of the particular solution with initial conditions $y = -.5$ and $y' = 3$ at $t = 0$. Although the graph shows dampening the repeating zeros of the equation show vibration.

APPENDIX

Complex numbers

The TI accommodates complex numbers in the form $a + bi$ where $i = \sqrt{-1}$. Be sure to use the special complex i and not the variable i. (2nd_i is above CATALOG.) The need for parentheses becomes acute when writing complex number operations. As an example in Figure A.1, the top two entries in the history area are the same sequence of numbers, but the placement of parentheses changes the meaning and result. The second screen shows that some functions allow complex number input. The entry ln(-1) normally gives an error, but using a complex number as input gives the result in complex terms.

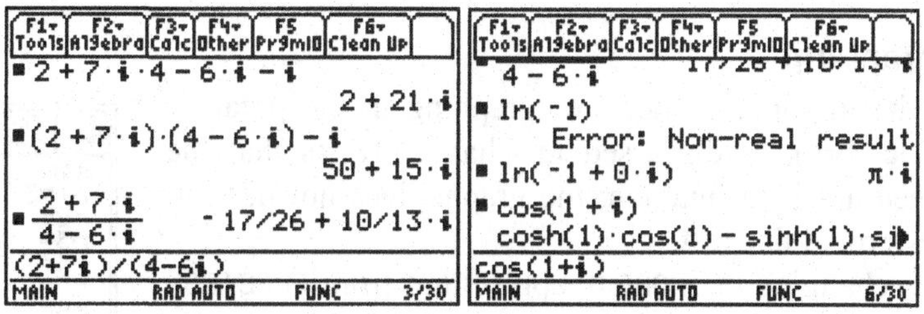

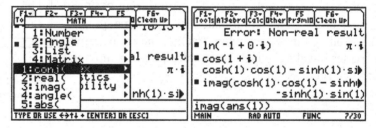

Figure A.1 Complex number arithmetic examples.

Complex operators

The 2nd_MATH, 5:Complex menu has a list of operations that can be used on complex numbers. This menu and a calculation are shown in Figure A.2.

Figure A.2 Using the 2nd_MATH, 5:Complex menu.

Polar coordinates in the complex plane

In MODE, the default Complex Format setting is Real, although Polar is an option. In most cases it is better to use the Real setting and convert numbers that need to be in polar format.

Coordinate conversion

Each point in the Cartesian rectangular coordinate system has a polar coordinate of the form (r, θ) where r is the distance to the origin and θ is a measure of rotation from the x-axis. The TI representation of polar form is $re^{i\theta}$.

In Figure A.3 we see the coordinate conversion tools ▶Polar which is found in the 2nd_CATALOG under P. The ▶Polar command puts the expression on its left into polar form. With Complex Format set to Real, any polar entry is converted to $a + bi$ format. However, there is also a ▶Rect command that puts the expression on its left into polar form. In addition, there are four commands that convert a representation in one form and give an individual component of the other form. This is done with the following commands that use functional notation:

    P▶Rx(rexpr,θexpr)
    P▶Ry(rexpr,θexpr)
    R▶Pr(xexpr,yexpr)
    R▶Pθ(xexpr,yexpr)

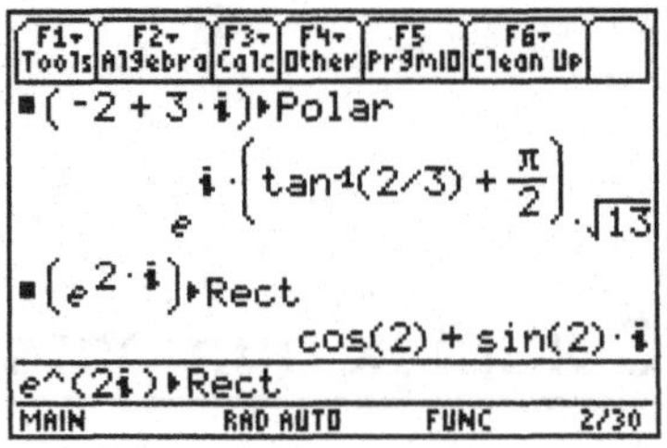

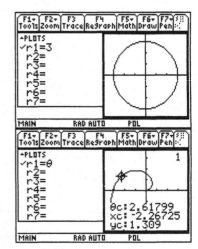

Figure A.3 Conversion techniques for complex numbers.

Tip: The theta (θ) key requires ♦ and is above the ^ key.

Polar graphing

One of the important aspects of graphing in the polar or parametric form, as discussed in Chapter 18, is that the curve need not be a function; the graphs does not need to pass the vertical line test.

After changing the MODE Graph setting from FUNCTION to POLAR, the Y= screen shows r= function definitions and it shows POL on the status line. In Figure A.4, a circle of radius 3 is drawn with a ZoomDec window.

In POL graphing mode, the independent variable is θ. Figure A.4 also shows the graph of $r = θ$ with the default coordinate setting (rectangular). Press ♦ _| to access the GRAPH FORMATS screen and change the setting from RECT to POLAR; and show polar coordinates of the same graph. Notice in both settings that, while tracing, the right arrow key increases θ-values which may correspond to moving any direction on the screen.

Figure A.4 Polar graph of a circle and spiral in a ZoomDec window.

3D graphing

We graph a function of one variable by thinking of the input values on the *x*-axis and the output values on the *y*-axis, i.e., setting $y = f(x)$. Functions can depend on more than one variable and in the case of a two variable function we can think of the inputs as *x* and *y* and set a third variable $z = f(x,y)$. The resulting points $(x, y, f(x,y))$ form a surface in three-dimensional space. The 3D graphing capabilities are useful in multivariable calculus allowing us to view surfaces with the 3D graph mode.

As an example, consider a function from economics called the Cobb-Douglas production function. It measures units produced based on two units, *x* for labor and *y* for capital. The general function is $f(x,y) = k \cdot x^a \cdot y^b$, where *a*, *b*, *k* are positive constants with $a + b = 1$. The normal sequence of graphing steps is required to produce a graph. See Figure A.5.

1. Press MODE and change the Graph setting to 3D.
2. Press ♦_Y= and enter the Cobb-Douglas function z1=10x^.6*y^.4.
3. Set the WINDOW. There are more settings to contend with and they require scrolling to see them all. The eye angles have to do with the viewing perspective of the surface. We change only the x, y, and z settings.
4. Press ♦_GRAPH and wait. A progress percentage is displayed in the upper left corner as it works.
5. Press F3 (Trace) to see productivity (zc) increases as x and y increase. The left and right arrows change the x-values, the up and down change the y-values. You can also enter the two specific numeric values, which is more convenient.

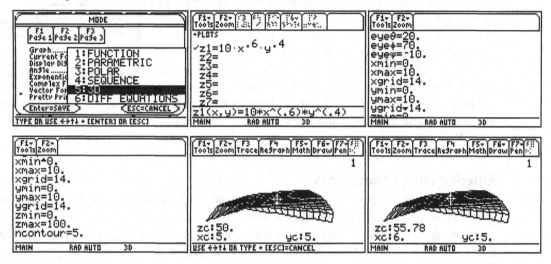

Figure A.5 The sequence of steps to draw a 3D graph.

Which has a greater impact on production, increasing labor or increasing capital? The answer depends upon a current position, so we consider (5, 5, 50) as our position. The last screen of Figure A.5 shows the point (6, 5, 55.78). You can also find that $x = 5$ and $y = 6$ gives $z = 53.7827$. This means that both increase productivity but an additional labor unit gives the greater productivity increase.

What inputs other than 5 and 5 give a production level of 50? This kind of question is answered by using contour levels. A contour level graph is two dimensional showing lines (curves) of point (x, y) that give the same z-value on the surface. It is like lines on a topographical map showing all the points of the same elevation. The TI draws contour levels.

Press ♦_| to bring up the GRAPH FORMATS screen. Set Style to 3:CONTOUR LEVELS and then press ENTER twice. We use the previous WINDOW settings; ncontour determines how many lines are drawn, and we use the default of 5.

After calculation, you see graph shown in Figure A.6 picture. On this screen, TRACE acts like the free-moving cursor – it does not follow contour lines. We find that $x = 7.14$ and $y = 2.86$ also give a production level near 50.

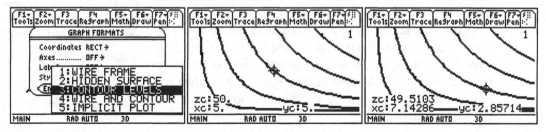

Figure A.6 Making a contour level graph from a 3D graph.

> *Tip:*   Once a 3D surface has been graphed, the eye settings for the window can be changed to see the surface from a new angle. This does not require surface recalculation time.

Using units

Applying calculus to physics has had spectacular success. A notable example is applying integration to find work done. The TI helps us handle the myriad of units used in physics and other sciences. It has a setting for each of the two main systems of units operating in the world, it is often necessary to make conversions between them.

The TI allows you to have one of three systems as the default conversion mode. This is selected in the MODE menu (Page 3). For example in SI mode all lengths are reported in meters, while in ENG/US the same length is reported in feet. You can even make up your own units using the CUSTOM mode; see the *TI Guidebook* for details.

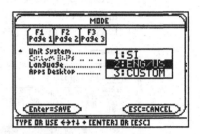

Figure A.7 The Unit System options in the MODE menu.

Entering units or constants

The key to understanding the TI system is to think of units (and constants) as multipliers that start with an underline. The underline symbol is the ♦ version of the MODE key.

How do we know feet are _ft, inches are _in and light-years are _ltyr? A complete menu of units and constants is activated by 2nd_UNITS and the three pages are shown in Figure A.8. Pressing the right arrow opens submenus. However, these are only the abbreviated names and you may need to consult the "List of Pre-Defined Constants and Units" section of *TI Guidebook* for full names.

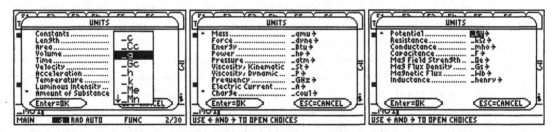

Figure A.8 Complete menus of units and constants (2nd_UNITS). Constants submenu shown.

The submenu list of constants is shown in Figure A.8. Because the constants cover such a range, they are hardest to guess from their symbol. One very common constant is the acceleration due to gravity, _g.

Constants are by default converted to the units of the current conversion mode. For example, in Figure A.9, when the conversion mode is ENG/US, then the gravity constant is shown in feet per second squared. But when the mode is SI, then the gravity constant is shown in meters per second squared.

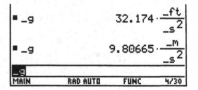

Figure A.9 The gravity constant in different systems. Between entries, we changed the Unit System from ENG/US to SI.

Length

As a simple example of conversion from one system to another, you may want to know how many kilometers you run in a marathon. The dictionary reports that it is 26 miles and 385 yards. We enter this as an addition in Figure A.10, and the answer is shown in feet because this is the default length in the ENG/US system. We want the answer in kilometers, so we add a convert symbol (▶) and the desired unit of length. The convert symbol is the 2nd version of the MODE key. This makes it easy to remember since the underline is the ♦ version of the MODE key.

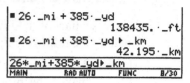

Figure A.10 The length of a marathon, with ENG/US system mode.

Caution about units

Suppose you wanted to know how many feet in a meter. Try entering: 1_m/1_ft. Figure A.11 shows an unexpected and incorrect answer in _ft². The history area shows that parentheses are needed to correct the input; a correct question insures a correct answer.

Applying units to a variable name requires that the name and the underline be separated by either a space or multiplication sign. Otherwise, the whole expression is assumed to be a single variable. When applying units to a number, the space or multiplication sign is optional, but the dot multiplication is applied and shows in all expressions in the history area.

If you wanted to know the area of a sine curved metal plate, $(0 < x < \pi)$ you might try putting units in an integration calculation. The bottom frame of Figure A.11 shows this attempt is met with an error message. In general, integration and differentiation do not allow the use of units.

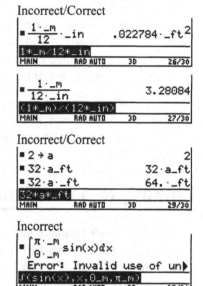

Figure A.11 Errors when using units.

Internet address information

The main internet address for Texas Instruments graphing calculator support is

http://www.education.ti.com/

At this site you use a search box or dropdown menus to find material of interest. This is the source for the *TI Guidebook*, activities, APPs and product information. To go directly to the *TI Guidebook* use:

http://education.ti.com/us/product/tech/89/guide/89guide1us.html

For discussion groups and an extensive downloadable program archive it is far better to use:

http://www.ticalc.org/

To go directly to the math program archive use:

http://www.ticalc.org/pub/89/basic/math/

Linking calculators

The essentials of linking are presented in the *TI Guidebook* and are not repeated here. But we include the following tips.

- The end-jack must be pushed firmly into the socket. The calculator turns on as it makes the proper connection. There is a final click you can feel as it makes the proper connection.

- If you are experiencing difficulty connecting, turn off both calculators, check the connection, and then turn them on and try again. If available, try other cables or calculators.

- When selecting items, use F4 (✓) to select and deselect, just as in the Y= editor.

- Folders can be expanded (opened) as an indented list or collapsed (closed) to just show the folder name. If the symbol ▶ follows the name, the folder is collapsed; it is expanded by pressing the right arrow. If the symbol ▼ follows the name, the folder is expanded; it is collapsed by pressing the left arrow.

- If you are required to drain your calculator memory before an exam, use F5 (All) 1:Select All to select everything to keep a copy on some other calculator. Even better is to store it on your computer, which is the next topic.

Linking to a computer

The TI-89 Titanium and Voyage 200 have a USB port connection and they come with a cable that connects to a USB port on a computer. The necessary software, TI Connect, is provided on a CD and also can be downloaded from the TI website. For the older TI-89 a TI-Graph Link™ package must be purchased separately, it is a cable and the TI Connect software. There is an older TI-Graph Link™.software program, but the newer TI Connect is more universal.

- This is the best way to back up your work.

- It is the preferred way to write and edit programs. Download a simulator or buy a TI keyboard if you write programs.

- You can download and transfer programs from the internet archives.

- It allows you to capture the screen in a form for direct printing or use in a word processor.

Troubleshooting

Nothing shows on the screen
- Check the contrast.
- Check ON/OFF button.

The screen is frozen
- If there is a busy indicator on the status line, then the TI may be still calculating. Press ON if you can't wait.
- If there is a pause indicator on the status line, then the TI is paused from a program. Press ENTER to continue.
- Press ESC several times to exit any menus and then press HOME to return to the HOME screen.

Y= Editor problems
- To edit a previous definition, you must first press either ENTER or F3 (Edit). Otherwise your keystrokes are ignored.
- Be sure to use the correct independent variable for the graph mode you are using (FUNC uses x, PARAMETRIC uses t, etc.).

Nothing shows up on the graph screen
- Press F3 (Trace) to see if a function is defined but has values outside the window.
- There may be no functions selected. This may be caused by using NewProb.
- The function may be graphed right along either axis; reset the window.
- If there is a busy indicator on the status line, then the TI may be still calculating. Press ON if you can't wait.
- If there is a pause indicator on the status line, then the TI is paused from a program. Press ENTER to continue.

Nothing shows up in the table
- You may have the Ask mode set and need to either enter x-values or change to Auto in the TBLSET.
- Check to see that a function is selected. (NewProb deselects them all.)

My changes don't take
- Options, modes and settings made within a dialog box must be confirmed for each item and then again for the whole box. Press ENTER twice after making changes.
- To edit a previous definition in the Y= editor, you must first press either ENTER or F3 (Edit). Otherwise your keystrokes will be ignored.
- Data cells in a table that have a formula in the header cannot be edited. Clear the formula.

Troubleshooting (continued)

I get an error message on the screen
This can cover the widest array of problems. Read the message carefully, and note where the cursor is positioned on the entry line. Press ESC and make a correction. If you have no idea what could have caused it, consult the Error message section of the *TI Guidebook* for explanations of the error messages. The following are quite common:
- Parenthesis mismatch. Count and match parentheses carefully.
- Pasting a command in the wrong place. For example, a program name must be on a fresh line entry line.
- Syntax errors. Too few or too many variables. Check the syntax with F1 in the CATALOG. In some cases, syntax differs, depending upon MODE settings.

I'm getting a result but it is wrong
In all cases, check the pretty print version in the history area first.
- Parenthesis mismatch. Count and match parentheses carefully.
- Subtraction vs. negative symbol. For example, the subtraction sign cannot be used to enter a -10.
- Order of operations. For example, 1/2x is x/2 on a TI and not the same as 1/(2·x).
- There may be an operation sign missing. For example, xy is a variable name, but x*y or x y is the product of x and y.
- Check that variable names you are expecting to be undefined have not been previously given a value. Use F6 (Clean Up) to clear variables.
- Check any applicable default settings.

Split screen confusion
Use 2nd-QUIT twice to insure getting back to a FULL screen; or press MODE and reset the options as desired.

Common problems with solve
- The output may be correct but not in the expected algebraic form.
- The first entry must be an equation, not an expression, as is the case with zeros.
- The equation entry is incorrect. You should check in the history area to see that it is what you meant.
- A coefficient may have been previously defined.
- An expression or variable may be a reserved word. See the *TI Guidebook* for a list of these.

My program won't run
Program errors are difficult to diagnose. Open the program code with the sequence APPS 7 2. Isolate the problem by adding temporary pauses and displays to check the progress.

INDEX